Coal Myths and Environmental Realities

Also of Interest

Fuels and Chemicals from Oilseeds: Technology and Policy Options, edited by Eugene B. Shultz, Jr., and Robert P. Morgan

Where We Agree: Report of the National Coal Policy Project, edited by Francis X. Murray

Decentralizing Energy Decisions: The Rebirth of Community Power, Ellis Cose

Transitions to Alternative Energy Systems, edited by Thomas Baumgartner

Equity and Energy: Rising Energy Prices and the Living Standards of Lower Income Americans, Mark N. Cooper, Theodore L. Sullivan, Susan Punnett, and Ellen Berman

Water and Western Energy: Impacts, Issues, and Choices, Steven C. Ballard, Michael C. Devine, and Associates

Coal Surface Mining: Impacts of Reclamation, edited by James E. Rowe

Mining Industry Permitting Guidelines: Coal Exploration and Production--The Western Region, Doann Houghton-Alico, Marcy Dunning, and Judith Goater

Energy, Economics, and the Environment: Conflicting Views of an Essential Interrelationship, edited by Herman E. Daly and Alvaro F. Umaña

Green Goals and Greenbacks: State-Level Environmental Review Programs and Their Associated Costs, Stuart L. Hart and Gordon A. Enk

+Science, Technology, and the Issues of the Eighties: Policy Outlook, edited by Albert H. Teich and Ray Thornton for the American Association for the Advancement of Science

+Available in hardcover and paperback.

Westview Special Studies in Natural Resources and Energy Management

Coal Myths and Environmental Realities: Industrial Fuel-Use Decisions in a Time of Change

Alvin L. Alm
with Joan P. Curhan

This book deals with two inconsistent myths that persistently surround industrial use of coal. The first myth is that the Clean Air Act precluded the use of coal; the second, that industrial use of coal will expand rapidly as a result of purely economic choices.

Through analyzing fuel-use decisions actually made by industry, Mr. Alm concludes that environmental quality standards have played a minor role in industrial choice of fuel. Historically, natural gas and oil have been both less costly and more convenient fuels for industry to use. Coal gained a substantial economic advantage over oil after the oil price increases of the last decade, yet it continues to maintain a lower market share than economics alone would suggest. Mr. Alm demonstrates that coal's share of the fuel market will continue to remain low because of the way U.S. businesses view fuel-use choices. For most U.S. firms, energy costs are a relatively small portion of total costs and a minor factor in a firm's ability to compete. Faced with alternative capital projects to expand production facilities or to make mandatory investments, companies do not generally give high priority to coal conversion projects. Moreover, most U.S. firms have little experience with burning coal, and that lack of expertise creates additional psychological and institutional barriers to coal's use. Finally, there is a tendency to prefer high-payoff, short-term investments over projects that promise cost savings many years in the future. These are all strong reasons for coal's lackluster performance in the industrial market--much more potent forces than environmental regulations.

Alvin L. Alm is now Deputy Administrator of the Environmental Protection Agency. This book was written when he was Director of Harvard University's Energy Security Program at the Energy and Environmental Policy Center and a lecturer at the John F. Kennedy School of Government at Harvard. Mr. Alm is co-editor of Oil Shock: Policy Response and Implementation (forthcoming).

Joan P. Curhan is Research Administrator at the Energy and Environmental Policy Center, Harvard University. Previously, she was Research Administrator of the Multinational Enterprise Project at the Harvard Business School. In connection with this project, she co-authored three books--Tracing the Multinationals (1977), The World's Multinational Enterprises (1973), and The Making of a Multinational Enterprise (1969).

A Report of the
Energy and Environmental Policy Center
Harvard University

Coal Myths and Environmental Realities: Industrial Fuel-Use Decisions in a Time of Change

Alvin L. Alm
with Joan P. Curhan

Westview Press / Boulder, Colorado

TO

MINNIE AND EMIL ALM
with deep appreciation

and

MURIEL AND DAVID POKROSS
with much love and thanks
for years of encouragement and caring

Westview Special Studies in Natural Resources and Energy Management

Published in 1984 in the United States of America by
Westview Press, Inc.
5500 Central Avenue
Boulder, Colorado 80301
Frederick A. Praeger, President and Publisher

Library of Congress Cataloging in Publication Data
Alm, Alvin L.
Coal myths and environmental realities.
(Westview special studies in natural resources and energy management).
Includes index.
1. Coal trade--Government policy--United States. 2. United States--Industries--Power supply--Decision making. 3. Fossil fuels--Environmental aspects--United States. 4. Fossil fuels--United States--Costs. 5. Air--Pollution--Law and legislation--United States.
I. Curhan, Joan P. II. Title. III. Series.
HD9546.A45 1984 333.8'22'0973 83-16940
ISBN 0-86531-712-7

Composition for this book was provided by the Energy and Environmental Policy Center.
Printed and bound in the United States of America.

5 4 3 2 1

Contents

Tables and Figures

Tables

Figures

Preface

This book was written while I was a lecturer and director of the Harvard Energy Security Program at the John F. Kennedy School of Government. Like any book prepared during a period of rapidly developing scientific investigation and evolving public policy concerns, the book is silent on the controversies surrounding the latest environmental issues. For example, I have not addressed here the impact of potential policy changes to cope with acid rain deposition. Rather the purpose of the research was to examine the role that environmental regulations have played to date in influencing industrial fuel-use decisions with special attention to the opportunities for increased utilization of coal. The examination of retrospective cases would provide an improved empirical foundation for evaluation of future policy proposals.

The book is short, but the research in pulling together ten case studies and conducting numerous interviews was extensive. Although we began with no particular bias or point of view, the conclusion--that environmental requirements have played a small role in industrial fuel-use decisionmaking--was surprising to the researchers who worked on the project. Indeed, the conclusion seemed contrary to the conventional wisdom. To assure accuracy, we twice sent copies of the report to all the industry participants. To satisfy myself on the report's conclusions, I talked to many friends in industry and presented the results to colleagues at Harvard and before other forums. My colleagues at Harvard urged that the book be published because they see in it a contribution to our understanding of industrial fuel-use decisionmaking processes.

This book represents the results of my research prior to becoming Deputy Administrator of the Environmental Protection Agency. Therefore, the book does not necessarily represent the position of the Environmental Protection Agency. It is the product of the work of a number of individuals, which I, as a member of the Harvard faculty, synthesized. I enjoyed the help of many, but I remain solely responsible for any remaining errors or omissions.

Alvin L. Alm

Acknowledgments

This book is the culmination of many months of hard work on the part of numerous students and staff at the Energy and Environmental Policy Center (EEPC) of the John F. Kennedy School of Government, Harvard University. I am particularly grateful to the EEPC, which not only provided me with a productive working environment, but also brought me into contact with many bright, talented, and enthusiastic students and staff, some of whom took on major responsibilities for portions of the book. Judith Hughes contributed as project director, supervising the case studies, overseeing the work of the student researchers, and in general, keeping the project moving forward. She also provided much of the substance and inspiration for the third chapter. Eugene Peters spent many months working on the analysis in Chapter 2, wrote the first draft of the chapter, and subsequently reworked the analysis when new data became available. Susan Smith wrote the first draft of Chapter 4, consulted on all parts of the manuscript concerning environmental law and regulations, and very generously gave of her time in updating the material in Chapter 4 and Appendix A after having left Cambridge for a full-time job in Washington, D.C. Karen Pierson assisted in the environmental analysis. Gary Anderson, Thomas Birdsall, William Brownell, Thomas Countryman, Loren Dunn, Chris Evangel, Carol Johnson, Tony Otten, Domonique van der Mensbrugghe, Mary Blakeslee, Robert Howe, and Brent Webber provided the core industry data in their thorough and well-researched case studies. Mark Bubriski, Chris Corpuz, Howard Feldman, Michael Long, Edward Edelson, and Ann Sachs conducted the interviews with industry managers that supplemented the material in the case studies. Robert Reinhard generously volunteered many hours to write summaries of the case studies and to edit portions of the manuscript. Scott Speck skillfully edited the first draft of the volume. Susan Elbow did an outstanding job of verifying the factual content of the manuscript and helped to pull the final pieces of the volume together. Cynthia Vissers O'Hanley did the final editing of the volume; because of her superb attention to detail, the end product is far more readable. Carol Donahue, with intelligence and unfailing good humor, made every correction and oversaw the final formatting of the manuscript. Susan Gould was most helpful in assisting with the final editing and metric conversions. At various points, Deborah Green, Chris Lunblad, Patricia Slote, and Leslie Sterling assisted with typing and editing. I am indebted to them all for their patience, good humor, and endurance as the manuscript went through draft after draft, correction upon correction, and migrated from word processing machine to word processing machine.

I am grateful to the company officials who kindly consented to be interviewed and who subsequently gave valuable time to reviewing the manuscript twice for accuracy. I am most grateful especially to William Hogan, Director of the Energy and Environmental Policy Center, and to my other colleagues at the Kennedy School--Henry Lee, Robert Leone, and David Harrison--who read and reread the manuscript, and gave me invaluable suggestions, comments, and encouragement when it was most needed. Finally, it is no understatement to say that without the professional expertise of Joan P. Curhan, this book would not have been published. Not only did she assist me in the early stages of data collection, but, upon my departure from Harvard, and on short notice, she oversaw all manner of details to meet deadlines.

The Andrew Mellon Foundation and the Environmental Protection Agency generously provided grant money without which I could not have undertaken the project and I wish to thank them both for their financial support.

Alvin L. Alm

Abbreviations

API - American Petroleum Institute

BACT - Best Available Control Technology

CEQ - Council on Environmental Quality

DOE - Department of Energy

EIA - Energy Information Administration
EOP - Emissions Offset Policy
EPA - Environmental Protection Agency
EPAA - Emergency Petroleum Allocation Act
EPCA - Energy Policy and Conservation Act
ESECA - Energy Supply and Environmental Coordination Act
ESP - electrostatic precipitators

FERC - Federal Energy Regulatory Commission
FPC - Federal Power Commission
FUA - Powerplant and Industrial Fuel Use Act

GNP - Gross National Product

HC - hydrocarbon

IRBs - Industrial Reserve Bonds
ITC - Investment Tax Credit

LAER - Lowest Achievable Emission Rate

NAAQS - National Ambient Air Quality Standards
NEES - New England Electric System
NESHAPs - National Emission Standards for Hazardous Air Pollutants
NGPA - Natural Gas Policy Act
NSPS - New Source Performance Standards

OECD - Organization for Economic Cooperation and Development

OPEC - Organization of the Petroleum Exporting Countries

PM - particulate matter

PSD - Prevention of Significant Deterioration

PURPA - Public Utilities Regulatory Policies Act

RCRA - Resource Conservation and Recovery Act

ROI - return on investment

SIPs - State Implementation Plans

TSP - total suspended particulates

Introduction

Two inconsistent myths persistently surround industrial use of coal. The first myth is that the Clean Air Act effectively precludes use of coal--whether through the high costs of compliance or through outright prohibitions. This concern is captured in the aphorism, "Coal is a wonderful fuel except you can't mine it or burn it." The second myth--or at least a major forecasting error--is that industrial use of coal will expand rapidly. Many future energy projections estimate industrial use of coal doubling or tripling over the next decade, even in the face of declining use since World War II. This study comes to opposite conclusions.

It is widely perceived that environmental standards cut coal use in the 1960s and made it increasingly difficult to use coal in the 1970s. Clearly, one major objection to the Clean Air Act has been its presumed chilling effect on coal use. In fact, there were perfectly logical economic reasons for coal's small and diminishing share of the industrial market--at least until the large world oil price increases resulting from the Iranian Revolution. Historically, the shift away from coal occurred because oil and natural gas were cheaper to use, more convenient, and cleaner. Even with today's higher oil and gas prices, which make coal much more economically attractive, environmental standards play a small role in industrial fuel-use decisions.

Our review of a large number of U.S. manufacturing firms found that environmental requirements killed only one project and seriously discouraged only one other--the first because the requirements made the project too costly, and the second because the firm feared delays in gaining a permit. By contrast, firms such as Du Pont, General Motors, and Ideal Cement pursued aggressive coal conversion programs, and other firms converted at least some of their plants. This does not mean that environmental controls are not costly, some times time-consuming, and certainly a nuisance to business; it does mean that they were not the major barrier to coal conversion.

If environmental requirements played such a minor role, why has industrial use of coal fallen so short of expectations? Some of the lag may be explained by the downturn in use of metallurgical coal--related to lower demand for steel, the declining economics of coke production, and changes in the mix of processes used for steel production. But this does not explain the drop in the amount of coal used for industrial boilers. While generally poor economics discouraged coal conversions, at least until the oil price spiral in 1979-80, there were certainly many more economic conversion candidates available than actual conversions. From this study, we found most conversions were initiated by some external event, such as unreliability of natural gas, expiring natural gas contracts, or new environmental requirements--<u>not by a hard look at the economics of using coal.</u>

Coal has had and will continue to have a low share of the industrial fuel market because of the way U.S. businesses view fuel-use choices. For most U.S. firms, energy costs are a relatively small portion of total costs and a minor factor in a firm's ability to compete. Faced with alternative capital projects to expand productive facilities or to make mandatory investments, coal conversion projects are generally not given high priority. Moreover, most U.S. firms have little experience with burning coal, creating additional psychological and institutional barriers to investments in coal conversion projects. Finally, firms tend to prefer high payoff, short-term investments to projects that promise cost savings many years in advance, particularly when those investments are unrelated to a firm's normal business operations. These are all strong reasons for coal's current lackluster performance in the industrial market. But there were different reasons in the past.

COAL ECONOMICS

Coal use began its downward spiral in the industrial sector after World War II. Coal was driven from industrial markets because its competitors--oil and natural gas--were cheaper and more convenient to use. Oil initially replaced coal in industrial use. As the natural gas pipeline system expanded outside producing areas, gas slowly became the dominant fuel for industrial operations. After 1966, when the oil import quota was lifted for residual fuel oil, oil became cheaper than coal <u>as a fuel</u> in the Northeast. Plants originally designed to burn coal were switched to oil, even though higher capital expenditures had been made to burn coal in the first place. Once conversion to oil was complete, states were able to tighten their emission standards, making it difficult to switch back to coal.

Despite strong national policy to increase coal use, coal actually lost ground after the Arab oil embargo. Because of an unusual coal price increase in 1974 and price controls on crude oil, coal's relative economic position declined. Since 1978, however, coal economics have changed dramatically. The Energy Tax Act of 1978 added a 10 percent investment tax credit to the one then in existence and eliminated the tax credit for oil or natural gas boilers, giving coal boilers a 20 percent investment tax credit advantage. Sparked by the Iranian interruption, the price of Saudi light crude rose from $12.75 per barrel to $34.00 per barrel by October 1981, raising the price of oil products and, indirectly, natural gas prices. The full impact of this large price increase was passed on to consumers when crude oil prices were finally decontrolled in early 1981. Industrial natural gas prices are now rising rapidly; soon they will be equivalent to residual fuel oil prices. Finally, the Economic Recovery Tax Act of 1981 allows five-year depreciation for boilers, which helps coal relative to other fuels since coal use requires much larger capital expenditures.

These changes have improved coal's economics measurably. Coal has economic advantages over oil in most boiler applications, and once industrial gas prices reach the price of oil substitutes, it will have similar advantages over natural gas. If economics were the only basis for industrial fuel-use decisions, coal's future would be brighter.

BEHAVIORAL CHARACTERISTICS

Classical economics clearly fails to describe the motivations for industrial firms to use coal. The gap between the economic analysis detailed in Chapter 2 and the reality of coal use can be explained by the behavioral and strategic impediments to coal use described in the industrial fuel-use model developed in Chapter 3. That model attempts to bridge the intellectual gap between modelers and those who are interested in the behavioral aspects of individual decisionmaking; it tries to explain the historical gap between econometric prediction and management behavior. The model describes four discrete stages in the firm's decisionmaking process, concentrating on those factors most important at each stage of the process. Based on many interviews and case studies, the model helps explain some of the most important conclusions reached in this study.

An industrial firm's decision whether to consider a certain facility for coal conversion is the single most important step in the process. Most facilities with characteristics favorable to coal use, such as large size or good transportation links, have never even been reviewed for their economic potential. Only a few firms systematically review their facilities for fuel switching potential. Unless firms have immediate competitive pressures to use coal, have high energy costs as a percentage of total costs, or have past favorable experience with coal, they are not likely to examine the possibilities for conversion without some form of external pressure. At this early stage, a decision to review fuel use entails a commitment of management time--time that, in some cases, is as precious as capital.

In most of the cases examined in this study, some external event--such as a plant expansion, an expiring natural gas contract, or a pollution compliance order--virtually forced the firm to review its fuel-use pattern. Only in the cement industry were economic forces inexorably pushing toward coal use. Because fuel use generally plays such a small part in the competitive relationships among firms, most economically viable coal conversion projects are never uncovered.

The firm's second critical decision point comes at the stage of capital budgeting. Since most coal conversions are large capital-intensive projects, managers must grapple with a range of strategic issues when a coal conversion project is at stake. They must weigh coal conversion projects against all other company priorities. They must allocate management time and attention to an activity outside the mainstream of their companies' business, one in which most firms have little or no experience. They must be willing to forego some productive investments or raise more debt and equity capital, often at the expense of diluting stock value and reducing credit ratings. And they must be willing to make commitments to retaining production at converted facilities for some time.

Despite these barriers, some conversions are indeed made. As noted, firms in the cement industry have been forced to convert by competitive pressures, and firms in the pulp and paper industry have felt some pressure to reduce oil and gas use. In a few cases where firms have developed experience using coal, they have been encouraged to be more aggressive in using fuel switching as a strategy to reduce energy costs. Other firms in the same industry without experience using coal would blanch at such a strategy.

Overall, the industrial fuel-use decisionmaking process is biased against coal. When interest rates are high and profits are low, manufacturing firms cannot afford the large capital expenditures that coal conversion projects require. When the level of economic activity expands, most firms rush to expand their production lines or develop new products--using their greater cash flow for "productive investments." In most industries, fuel costs do not represent a sufficiently large percentage of total costs to induce companies to allocate large amounts of capital and management time to coal conversion. Without strong competitive pressures to reduce fuel costs--beyond what can be achieved by conservation--industrial coal use will fall short of expectations.

ENVIRONMENTAL BARRIERS

Although environmental requirements are not the primary obstacle to greater industrial use of coal, they do raise costs, cause delays, and add to the uncertainty of coal use. In the past, environmental costs have been relatively small; electrostatic precipitators, for example, add about 10 percent to the capital costs of using coal but less than 5 percent to the total costs. If stack gas scrubbers are required, environmental costs would make up a third of capital costs and result in a substantial increase in operating costs for a moderate-sized boiler. Currently, most industrial users meet environmental emission standards through the use of low sulfur coal, which they purchase at a small premium.

Permit delays were not a major factor in determining fuel-use choices. In the majority of cases, permits were received within planned time periods, although for some firms, the period could be as long as three years if no previous monitoring had been conducted. For many firms, delays would cost more money and management time, but would have no other serious impacts. Delays were a more serious problem in highly competitive industries, however, where success often depends on speed and secrecy. In these industries, a protracted permitting process gives competitors an opportunity to develop countervailing strategies. Recognizing this threat, some firms have eliminated potential coal conversion sites.

Uncertainty was not nearly as serious a barrier to coal conversion as often touted. Most industry environmental staffs are well informed about what is expected from regulatory agencies. This is particularly true in the South, where most potential for coal conversion exists and in which environmental agencies are generally accommodating. Firms expressing most uncertainty were usually those with little experience using coal. One firm with extensive experience actually sped up a project to prevent potential future shifts in government policy; it assumed that once the permit was received, it would be grandfathered against future regulatory changes.

Environmental requirements could be more limiting in the future. Clean Air Act limits, designed to protect air quality in areas of the country not violating ambient air quality standards, could become binding--not only on coal conversion--but in some limited cases, on economic development in general. These Prevention of Significant Deterioration (PSD) limits could be particularly troublesome in the Southwest, with its massive concentration of industry, and in hilly terrain, where difficult meteorological conditions prevail. But no one knows how serious an impediment PSD will become until detailed studies are conducted at specific sites. In some cases, a modest change in location or reduction of pollution from other sources at the plant would allow coal to be burned. For those cases where coal simply cannot be burned and alternative sites or offsets are not available, natural gas will generally be available. Since so much industrial growth is occurring in the South and Southwest, which are also major natural gas producing areas, oil will generally not be the option chosen.

METHODOLOGY

This study approached the impact of environmental requirements on industrial fuel-use decisions from a number of angles. The first chapter looks at the history of industrial energy use in the United States, focusing on the period from 1951 to 1981. It also explores the role of government policy in fuel-use decisions, particularly since 1973. Chapter 2 reviews the economics of coal and oil use in boilers--before and after the Iranian Revolution and with and without stack gas scrubbers. The third chapter reviews the factors that influence industrial fuel-use decisions, developing a model of corporate fuel-use decision-making. In many respects, this chapter provides the most original insights from this study. The fourth chapter reviews how environmental factors influence fuel-use decisions and the options available to industry for overcoming potential environmental restrictions. The final chapter reviews the importance of coal use and conversion as an overall part of national energy policy and suggests ways to overcome barriers to coal use.

The study employs case studies as the basic methodological approach. (See Appendix B for a brief description of the case studies.) Detailed case studies of ten firms were buttressed with interviews with a large number of industry officials. In addition, a mathematical model was employed to evaluate the economics of using coal or residual fuel oil.

The study's focus on coal conversions was chosen for a number of important reasons. First, conversions offer great potential for reducing oil and gas use; industrial boilers use about 10 percent of U.S. primary energy. Second, although coal will often be attractive for new facilities, it is not clear that many "greenfield" manufacturing facilities will actually be built in the United States for some time--at least in major energy-using industries. Most basic U.S. manufacturing industries face stagnant demand and excess capacity. Even in expanding industries, it is likely that most growth will occur at existing sites. Since most firms currently have spare steam capacity, they can choose to meet part of their steam requirements from spare capacity and the rest from new oil or gas boilers. In a sense, many decisions on fuel use for expanding facilities resemble a conversion decision. The firm must decide to scrap existing oil and natural gas boilers to make way for a large coal-fired boiler. Because of the nature of these decisions, this report focuses on conversions from oil and gas to coal (and sometimes wood) in which oil and gas boilers would be retired.

1
History of Industrial Fuel Use

Despite momentous changes in the world oil and domestic gas markets, U.S. industrial coal use has fallen since World War II. It has declined over 60 percent since 1951, over a third of that decline having occurred since the Arab oil embargo. Coal's position as an industrial fuel continued to deteriorate through the 1970s, despite the peaking of domestic natural gas and oil production in the early 1970s, two oil price spikes from supply interruptions in 1973 and 1979, and substantially higher prices for both oil and natural gas.

To understand why coal use has historically fallen so short of expectations, one must look at how the industrial fuel mix has changed over time and the role government policies have played in pricing different fuels. The analysis begins with pre-20th century fuel-use patterns, but quickly advances to two more recent periods. The first, from 1951 to 1973, was a time of high economic and energy growth, during which real energy prices fell constantly, oil demand increased by 70 percent, and natural gas demand more than tripled. The second, from 1973 to 1981, was characterized by two oil price shocks, generally sluggish economic growth, and almost complete government domination of fuel use. By looking at energy demand patterns over these time periods, we can at least gain clues about the influence of environmental standards on fuel use.

THE EARLY YEARS

Compared to Europe, the United States was a latecomer to coal. Not until the middle of the 19th century did coal's cost advantages and flexibility become obvious. The greatest initial demand came from the iron industry, whose entire works previously had to be moved whenever the wood supply was depleted. Since a half ton (.4535 metric tons) of coal could replace two tons (1.814 metric tons) of wood at half the cost, coal not only reduced costs substantially but also allowed for construction of permanent facilities.[1] Coal use spread rapidly, from 10 percent of all fuel in 1850 to 77 percent in 1910.[2] From that point on, however, coal almost consistently lost total market share.

Coal declined for precisely the same reasons that it had replaced wood in the first place: Alternative fuels became cheaper and more convenient to use. After World War I, oil began to replace coal in homes, office buildings, ships, and railroads. Although the decline in the industrial sector was less marked, coal's importance relative to total industrial output was reduced sharply. In the manufacturing industries, coal use for heat and power increased only 7 percent between 1909 and 1954 despite a fivefold increase in production, a twelvefold increase in fuel oil consumption, and a twenty-threefold expansion in natural gas use.[3]

Oil was a logical competitor in the industrial market since it was cheap, did not require the rail and barge infrastructure of coal, and could be stored easily. Oil prices were generally low during the early parts of the 20th century, as new finds drove down prices. Even after John D. Rockefeller established "price discipline" in the domestic market and the Texas Railroad Commission later imposed conservation restrictions, the "oil problem" was always perceived by the industry to be low and erratic prices. Even with domestic prorationing and international pricing agreements, oil prices were still relatively low, leading to substantial demand increases. By 1951, oil provided 23 percent of industrial fuel use and by 1973 it provided 28 percent of industrial fuel use.[4]

Prior to the development of the welded pipeline, natural gas was used only near producing areas. Even as late as 1945 most natural gas was still used in the state where it was produced. Of the 6 trillion cubic feet (tcf) (168 billion cubic meters) produced in 1950, only 1 tcf (28 billion cubic meters) of marketed production was transported in interstate commerce. But by the mid-1950s some 5 tcf (140 billion cubic meters) went into interstate commerce.[5] Once natural gas availability spread, gas became the dominant industrial fuel since it was both cheaper and cleaner than its competitors. By 1951, natural gas constituted 22 percent of the industrial market and by 1973, 32 percent of the industrial market.[6] In general, only a lack of pipelines hampered more widespread use.

1951-1973: CONTINUED RISE OF OIL AND GAS

Industrial coal use began a steep decline in the 1950s. Although not the focus of this study, industrial use of coking coal fell by 17 percent between 1951 and 1973. (Coke use rises and falls on the basis of total steel demand, processes employed, and import levels.) The more dramatic drop came from coal use for industrial boilers and process uses, which is the focus of this study. From 1951 to 1973, coal uses for these purposes fell by 47 percent, as shown under the column "Other Industry and Miscellaneous" in Table 1.1. This slackening of demand led some to conclude that environmental requirements caused this reduction in coal use.[7]

Poor economics, however, rather than environmental restrictions were responsible for coal's decline. Table 1.2, which lists average prices for domestic fuels between 1951 and 1981, shows that, on average, natural gas was always cheaper than coal until 1979. Hence, as the interstate pipeline system expanded, industrial firms turned to natural gas whenever they could. By 1974, natural gas constituted 32 percent of the industrial market.[8] Most striking, it captured 60 percent of the large boiler market--the only market in which coal was competitive at that time (see Table 1.3). Coal held only 17 percent of that market by 1974.

During the 1951-73 period, oil was generally cheaper than coal for industry to use except in those cases where industries had extremely large boilers with high capacity factors. Coal is cheaper to use in larger boilers because the ratio of fixed costs to total costs decreases as size increases. It is also cheaper in high capacity operations since the fixed costs are spread across many more energy units. The fact that oil cost almost twice as much as coal on an equivalent energy basis did not generally compensate for the higher capital costs associated with coal use.

During the 1960s, New England was coal's Waterloo. Coal had already been losing its New England market in the mid-1960s, but by 1966 it was actually more expensive than oil <u>as a fuel</u>--without even taking into account coal's higher capital and operating costs (see Table 1.4). The lifting of the mandatory import quota on residual fuel oil in March 1966 resulted in substantial reductions in East Coast residual oil prices, while coal prices rose slightly and gas was constrained by pipeline capacity. (Removal of the oil import quota on crude oil did not occur until 1973.) This unusual price relationship led to conversions to oil throughout the Northeastern states, even in large utility boilers where coal had the strongest relative position.[9]

Because all fossil fuels were relatively cheap prior to 1973, transportation played a larger role in determining costs to end-users. Natural gas, for

TABLE 1.1

COAL CONSUMPTION BY END-USE SECTOR, 1951-1981
(million short tons)

Year	Electric Utilities	Industry and Miscellaneous: Coke Plants	Industry and Miscellaneous: Other Industry and Miscellaneous	Industry and Miscellaneous: Total	Transportation	Residential and Commercial	Total
1951	105.8	113.7	128.7	242.4	56.2	101.5	505.9
1952	107.1	97.8	117.1	214.9	39.8	92.3	454.1
1953	115.9	113.1	117.0	230.1	29.6	79.2	454.8
1954	118.4	85.6	98.2	183.9	18.6	69.1	389.9
1955	143.8	107.7	110.1	217.8	17.0	68.4	447.0
1956	158.3	106.3	114.3	220.6	13.8	64.2	456.9
1957	160.8	108.4	106.5	214.9	9.8	49.0	434.5
1958	155.7	76.8	100.5	177.4	4.7	47.9	385.7
1959	168.4	79.6	92.7	172.3	3.6	40.8	385.1
1960	176.6	81.4	96.0	177.4	3.0	40.9	398.0
1961	182.1	74.2	95.9	170.1	0.8	37.3	390.3
1962	193.2	74.7	97.1	171.7	0.7	36.5	402.2
1963	211.3	78.1	101.9	180.0	0.7	31.5	423.5
1964	225.4	89.2	103.1	192.4	0.7	27.2	445.7
1965	244.8	95.3	105.6	200.8	0.7	25.7	472.0
1966	266.5	96.4	108.7	205.1	0.6	25.6	497.7
1967	274.2	92.8	101.8	194.6	0.5	22.1	491.4
1968	297.8	91.3	100.4	191.6	0.4	20.0	509.8
1969	310.6	93.4	93.1	186.6	0.3	18.9	516.4
1970	320.2	96.5	90.2	186.6	0.3	16.1	523.2
1971	327.3	83.2	75.6	158.9	0.2	15.2	501.6
1972	351.8	87.7	72.9	160.6	0.2	11.7	524.3
1973	389.2	94.1	68.0	162.1	0.1	11.1	562.6
1974	391.8	90.2	64.9	155.1	0.1	11.4	558.4
1975	406.0	83.6	63.6	147.2	*	9.4	562.6
1976	448.4	84.7	61.8	146.5	*	8.9	603.8
1977	477.1	77.7	61.5	139.2	*	9.0	625.3
1978	481.2	71.4	63.1	134.5	*	9.5	625.2
1979	527.1	77.4	67.7	145.1	*	8.4	680.5
1980	569.3	66.7	60.3	127.0	*	6.5	702.7
1981■	596.2	60.8	64.4	125.1	*	6.4	727.7

*Less than 0.05 million short tons. Quantities are included in the Other Industry and Miscellaneous category.
■Preliminary. Note: Sum of components may not equal total due to independent rounding.

Sources: 1951 through 1975--Bureau of Mines, Minerals Yearbook, "Coal--Bituminous and Lignite and Coal--Pennsylvania Anthracite" chapters. 1976--Energy Information Administration, Energy Data Report, Coal--Bituminous and Lignite in 1976 and Coal--Pennsylvania Anthracite 1976. 1977 and 1978--Energy Information Administration, Energy Data Report, Bituminous Coal and Lignite Production and Mine Operations--1977;...1978 and Coal--Pennsylvania Anthracite 1977;...1978. 1979 through 1981--Energy Information Administration, Energy Data Report, Weekly Coal Report. Department of Energy, Energy Information Agency, 1981 Annual Report to Congress, Vol. 2, May 1982, p. 127.

TABLE 1.2

PRICES OF DOMESTICALLY PRODUCED FOSSIL FUELS, 1951-1981
(cents per million Btu)

Year	Crude Oil[1]		Natural Gas[2]		Bituminous Coal and Lignite		Anthracite		Composite[3]	
	Current	Constant[4]	Current	Constant[4]	Current	Constant[4]	Current	Constant[4]	Current	Constant[4]
1951	43.6	76.4	6.6	11.6	18.8	32.9	40.2	70.4	25.5	44.7
1952	43.6	75.3	7.0	12.1	18.7	32.3	38.9	67.2	25.7	44.4
1953	46.2	78.5	8.2	13.9	18.8	32.0	40.2	68.3	26.9	45.7
1954	47.9	80.4	9.1	15.3	17.3	29.1	35.6	59.8	27.4	46.0
1955	47.8	78.6	9.3	15.3	17.3	28.4	32.6	53.6	27.0	44.4
1956	48.1	76.6	9.7	15.4	18.6	29.6	34.2	54.5	27.5	43.8
1957	53.3	82.1	10.2	15.7	19.6	30.2	37.6	57.9	29.7	45.7
1958	51.9	78.6	10.7	16.2	18.7	28.3	37.3	56.5	28.9	43.8
1959	50.0	74.0	11.6	17.2	18.6	27.5	35.1	51.9	28.3	41.9
1960	49.7	72.3	12.6	18.3	18.3	26.6	33.0	48.0	28.1	40.9
1961	49.8	71.8	13.6	19.6	17.9	25.8	33.8	48.8	28.5	41.1
1962	50.0	70.8	14.0	19.8	17.5	24.8	32.8	46.5	28.4	40.2
1963	49.8	69.5	14.3	20.0	17.2	24.0	35.7	49.8	28.0	39.1
1964	49.7	68.3	14.0	19.2	17.5	24.0	37.0	50.8	27.7	38.1
1965	49.3	66.3	14.2	19.1	17.5	23.5	35.3	47.5	27.5	37.0
1966	49.7	64.7	14.2	18.5	17.9	23.3	33.7	43.9	27.7	36.1
1967	50.3	63.6	14.5	18.3	18.4	23.3	34.7	43.9	28.3	35.8
1968	50.7	61.4	14.7	17.8	18.6	22.5	37.6	45.6	28.5	34.5
1969	53.3	61.4	15.1	17.4	20.0	23.0	42.3	48.7	29.6	34.1
1970	54.8	59.9	15.5	16.9	25.5	27.9	47.1	51.5	31.6	34.6
1971	58.4	60.8	16.5	17.2	29.2	30.4	51.4	53.5	33.9	35.3
1972	58.4	58.4	16.9	16.9	31.9	31.9	52.9	52.9	34.6	34.6
1973	67.1	63.5	19.8	18.7	35.5	33.6	58.9	55.7	39.4	37.3
1974	118.4	103.0	27.7	24.1	66.4	57.8	98.4	85.6	67.3	58.6
1975	132.2	105.3	40.6	32.3	82.9	66.0	137.9	109.8	82.0	65.3
1976	141.2	106.9	53.1	40.2	83.9	63.5	149.0	112.8	89.8	68.0
1977	147.8	105.7	72.3	51.7	87.3	62.4	150.4	107.6	100.6	71.9
1978	155.2	103.4	83.2	55.4	97.1	64.7	149.9	99.9	111.1	74.0
1979	217.9	133.9	107.9	66.3	104.7	64.3	174.1	107.0	141.3	86.8
1980	365.3	206.0	145.9	82.3	106.1	59.8	188.4	106.2	200.2	112.9
1981[5]	535.0	277.6	187.7	97.4	112.3	58.3	203.4	105.6	270.4	140.3

[1] Includes lease condensate.

[2] Wet natural gas, prior to extraction of natural gas plant liquids.

[3] Derived by multiplying the price per Btu of each fossil fuel by the total Btu content of the production of each fossil fuel and dividing the accumulated price of total fossil fuel production by the accumulated Btu content of total fossil fuel production.

[4] Constant 1972 prices calculated using GNP implicit price deflators, 1972 = 100.

[5] Estimated.

Note: All fuel prices taken as close as possible to the point of production.

Source: Department of Energy, Energy Information Administration, 1981 Annual Report to Congress, Vol. 2, May 1982, p. 21.

TABLE 1.3

INDUSTRIAL FUEL USE (BEST ESTIMATE CASE)

	Historical 1974[1]			Estimates 1979[2]		
	Quads/ Year	MMB DOE	Fractional Shares[3]	Quads/ Year	MMB DOE	Fractional Shares[3]
MFBI (Boilers Larger than 100 mmBtu/hr of Input):						
Oil	.5	.3	.13	.6	.3	.13
Gas	2.5	1.2	.63	2.8	1.3	.61
Coal	.7	.3	.17	.8	.4	.17
Renewables	.3	.1	.07	.4	.2	.09
Subtotal	4.0	1.9	1.00	4.6	2.2	1.00
Small Boilers and Process Heaters:						
Oil	2.3	1.1	.25	3.6	1.8	.34
Gas	4.7	2.2	.52	4.9	2.3	.46
Coal	.8	.4	.09	.8	.4	.08
Electricity	.1	.1	.01	.1	.1	.01
Renewables	1.2	.6	.13	1.2	.6	.11
Subtotal	9.1	4.4	1.00	10.6	5.2	1.00
Feedstocks:						
Oil	2.8	1.6	.50	3.2	1.7	.52
Gas	.5	.2	.09	0.9	.4	.14
Met Coal	2.3	1.1	.41	2.1	1.0	.34
Subtotal	5.6	2.9	1.00	6.2	3.0	1.00
Other and Unspecified Uses:						
Electricity[4]	2.3	1.1	.85	2.8	1.3	1.00
Other	.4	.2	.15	-	-	-
Subtotal	2.7	1.3	1.00	2.8	1.2	1.00
Fuels Consumption Totals:						
Oil	5.6	3.0	.29	7.4	3.8	.31
Gas	7.7	3.6	.34	8.6	4.1[5]	.36
Coal	3.8	1.8	.17	3.7	1.7[5]	.15
Electricity	2.4	1.2	.11	2.9	1.4	.12
Renewables	1.5	.7	.07	1.6	.8	.06
Other	.4	.2	.02	-	-	-
GRAND TOTAL	21.4	10.5	1.00	24.2	11.8	1.00

[1] Energy Information Administration (EIA), Energy Consumption Data Base with adjustments.

[2] EIA, September 1980, Monthly Energy Review, with the estimated renewable contribution added in.

[3] Based on the Quads/Year column.

[4] Primarily machine drive and electrolytic electricity uses.

[5] Totals do not add, due to rounding.

Source: U.S. Department of Energy, Reducing U.S. Oil Vulnerability: Energy Policy for the 1980s, Nov. 10, 1980, p. IV-H-18.

TABLE 1.4

ANALYSIS OF FUEL COST FOR ELECTRIC GENERATION,
NEW ENGLAND
(cents per million Btu)

	Coal	Oil	Natural Gas
1964	33.6	34.5	34.0
1965	33.6	34.4	33.5
1966	33.8	33.2	33.5
1967	34.2	30.9	31.7
1968	34.3	29.9	31.5
1969	34.7	29.9	33.3
1970	34.9	35.6	31.2

Source: Edison Electric Institute, *Historical Statistics of the Electric Utility Industry Through 1970* (EEI Publication No. 73-34), Table 42, p. 116.

example, was just as expensive as oil in New England because of the high costs of transporting gas to that region. Indeed, transportation costs dictated regional fuel preferences across the nation. Coal continued to dominate the Mid-Atlantic states and had a strong foothold in the Great Lakes states because of their proximity to coal fields. Oil dominated in New England. Natural gas dominated the industrial market in gas producing states, the West, Midwest and Southeast. By 1973, natural gas was the major industrial fuel in seven out of the ten federal regions; indeed, it dominated all markets unless transportation costs gave coal or oil an advantage.

Despite this strong showing by natural gas, the transition away from coal was slow. The average decrease in coal use over the twenty-two year period was only about 2 percent annually. Much of the shift occurred as industry moved to the Sunbelt, as new plants were built, or as old ones were renovated. This data does not suggest any wholesale shifting from coal to oil or natural gas at facilities then burning coal, except during the mid-1960s in New England. This slow transition from a fuel source that was often more costly, dirty, and bulky to fuel sources that were often cheaper, cleaner, and more flexible suggests that a movement back to coal will be much slower than many predict.

This twenty-two year historical snapshot suggests that environmental requirements were not important in shaping fuel-use decisions. Natural gas use grew fast because it was cheap, particularly when used close to the producing areas. Oil was always important in the Northeast and that importance grew when it had an absolute cost advantage over coal. Coal held markets chiefly in the regions closest to the coal fields. Pervasive economic forces shaped fuel-use patterns in the United States over this period of time; environmental requirements played only a minimal role.

1973-1981: THE PRIMACY OF GOVERNMENT POLICY

The 1973-81 period is noteworthy for its turmoil in energy markets. Two oil supply interruptions, a coal strike, a fivefold increase in real world oil prices, constant curtailments of natural gas supplies, and one serious winter natural gas shortage all took place during this tumultuous period. Even earlier, four major events had irreversibly changed the U.S. energy future. U.S. production of crude oil peaked in 1970; natural gas production peaked in 1973; marginal costs for new electric generating facilities rose above average costs in 1969, after decades of declining costs; and the OPEC countries consolidated market power during the early years of this decade.

The Arab oil embargo defined the "energy crisis" to most Americans. Although the producing nations were already gaining market power through individual bargaining arrangements initiated by Colonel Moammer Qadhafi of Libya and the Shah of Iran, it took the embargo to reshape oil prices and bring home the full extent of Western vulnerability to supply interruptions. From October 1973 to the beginning of 1974, the price of Saudi Arabian marker crude oil rose from under $3 per barrel to $11.65 per barrel. This dramatic increase, however, was partly sheltered from domestic consumers by imposition of price controls.

These higher prices should have resulted in a reduction in oil use, as occurred in all the other major industrialized countries except Canada. But the combination of price controls on oil, curtailments of natural gas supplies and a large coal price hike led to greater oil use, despite clear national policy to the contrary. Lower demand caused a three-year decline in total energy consumption and a five-year decline in industrial energy use. But, surprisingly, industrial oil use expanded throughout this period.[10] Industrial demand for refined petroleum products increased 19 percent from 4.5 million barrels per day (mbd) in 1973 to 5.35 million barrels per day (mbd) in 1979.[11]

Figure 1.1 shows that the ratio of domestic crude oil prices to coal prices deteriorated from the time of the Arab oil embargo until the Iranian Revolution. Even though oil price controls prevented U.S. oil prices from reaching the prevailing world market prices, one would have expected oil prices to rise faster than coal prices. But for a variety of reasons the opposite actually occurred. In 1969 and 1970, demand for coal by utilities, domestic steel companies and export customers rose, resulting in pressures on prices. At the same time, the Federal Coal Mine Health and Safety Act was being implemented, decreasing underground mining productivity and increasing costs. The industry was rapidly expanding and bringing in many new, untrained workers. Unsettled labor conditions generated concerns that the United Mine Workers might strike in November 1974, prompting customers to build up coal inventories to meet peak winter demands. When the Arab oil embargo hit, the coal market was already tight and prices were rising. Coal prices exploded after the embargo, leaping 72 percent in 1974, while domestic oil prices, contained by government price controls, increased only 62 percent during the same year (see Table 1.2).

Once oil and coal prices rose so dramatically, one would have expected natural gas demand to pick up measurably. Average natural gas prices were only a fourth of those of oil and 40 percent of coal costs, so

FIGURE 1.1

RATIO OF AVERAGE CRUDE OIL PRICES
TO COAL PRICES*

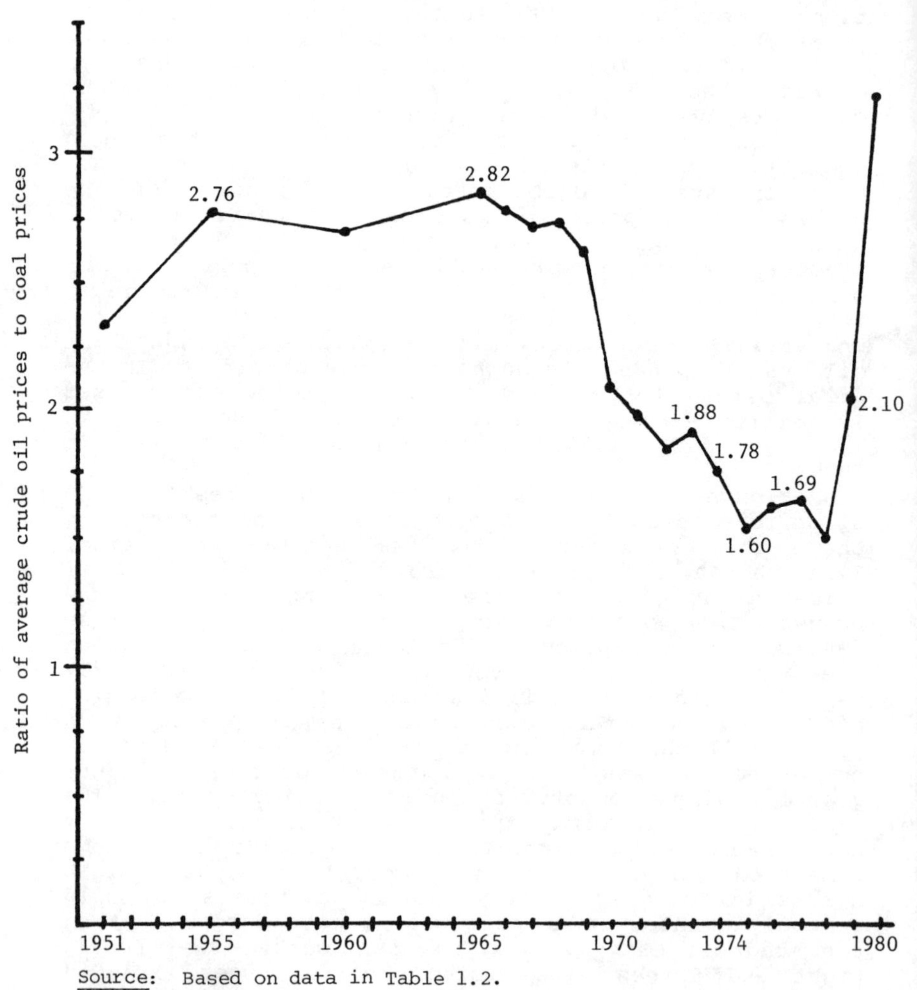

Source: Based on data in Table 1.2.

*Coal prices at minemouth. Oil prices at the wellhead.

there was plenty of room for gas price increases without offsetting the latter's competitive price edge. Had natural gas wellhead prices never been controlled, the higher prices would have stimulated greater exploration and development activity and ultimately production of new sources of natural gas. But natural gas price controls resulted in a 2.5 tcf (70 billion cubic meters) drop-off in total production between 1973 and 1975. Most of that reduction--1.8 tcf (50.4 billion cubic meters)--consisted of curtailment to industrial customers.

During the 1973-81 period, it was government policy, not the marketplace, which dominated energy supply and demand. Price controls on oil and natural gas overstimulated demand and held back supply. For natural gas, the price control system caused a contraction of production, which in turn led to a large curtailment of supplies available to the industrial market. Price controls on domestic oil also stifled production, but induced greater demand and hence oil imports. Only coal prices were unregulated, but every other aspect of coal mining, transportation, and use were subject to government control. To understand how government shaped the energy market, and particularly the industrial market, it is useful to examine the genesis of these policies and how they actually worked during this period.

Natural Gas

As pipelines created a national market for natural gas, the pressure for federal regulation grew. While state commissions could regulate the profits of distributors within a state, they had little control over wellhead prices or the profits of pipelines. When Appalachian gas production was replaced with more costly gas from other states in the mid-1930s, the political situation was ripe for legislation controlling natural gas prices.

The Natural Gas Act of 1938 enabled the Federal Power Commission (FPC), created by the National Power Act a few years earlier, to regulate the price that natural gas pipelines charged distributors. The FPC did not interpret its mandate to regulate the price at which producers sold natural gas to pipelines, but rather to regulate the rate of return for interstate pipelines. However, in 1954 the Supreme Court concluded in Phillips Petroleum vs. Wisconsin that the FPC was obligated to exert control over the inputs distributed through interstate pipelines. In other words, the FPC was to regulate the wellhead price of natural gas. The ink was hardly dry on the Phillips decision when Congress prepared decontrol legislation. Backed by the Eisenhower Administration, Congress passed legislation in 1956, only to have it vetoed by President Eisenhower because of the publicity surrounding a bribe offered to a U.S. Senator by a decontrol lobbyist.[12]

FPC regulation of interstate natural gas prices resulted in the creation of two distinct markets. Prices for natural gas sold outside the state in which it was produced were regulated by the FPC, while prices for intrastate supplies were unregulated. Interstate prices, established on the basis of production costs rather than on the value of the commodity, began to lag behind intrastate prices substantially by the early 1970s (see Table 1.5). As long as gas supplies were adequate to meet demand at these controlled prices, the dual market posed no apparent problem. Once natural gas production peaked, however, most new additions were funneled to the more profitable intrastate market, causing serious curtailments and switching to oil in the interstate market.

Declining reserves and production of natural gas led the next three administrations to seek a different regulatory system. The Nixon and Ford administrations both unsuccessfully proposed natural gas decontrol legislation. While Congress failed to act, the Federal Power Commission increased new interstate gas prices by over 100 percent from 1974 to 1977.

In 1977, President Carter requested legislation to tie natural gas prices to the average price of crude oil--getting rid of the existing cost-based regulatory system. Since crude oil prices would continue to be controlled, natural gas prices would not have reached full market prices. The House passed legislation similar to this proposal, but the Senate rejected it, passing instead a bill providing for phased decontrol of natural gas prices. After more than a painful year in a House-Senate conference, the Natural Gas Policy Act (NGPA) was born. The NGPA raised the prices for many different categories of natural gas, extended controls to the intrastate market, and created a schedule for phasing out natural gas price controls for new sources of supply. While this ungainly legislative instrument won little praise, it reduced natural gas shortages, increased reserve additions, and eliminated the prevailing dual market system.

Government price controls on natural gas first created excessive demand for gas in the industrial sector and then truncated the supply. In 1950, natural gas at the wellhead was eight times cheaper than crude oil; in 1960, it was four times cheaper; and by 1970 it was 3.5 times cheaper.[13] Today natural gas prices are fast approaching oil prices, with some distributors losing industrial customers to residual fuel oil. Because of the large price gaps of the past, it is no wonder that industrial users were drawn to natural gas

TABLE 1.5

U.S. NATURAL GAS PRICES, 1966-1977
(dollars per thousand cubic feet)

	Interstate		Intrastate
	Average Price	Net Contract Price	New Contract Price
1966	$0.17	$0.17	$0.20*
1967	0.17	0.19	0.22*
1968	0.17	0.19	0.23*
1969	0.18	0.20	0.26*
1970	0.18	0.20	0.26*
1971	0.18	n.a.	0.29*
1972	0.19	0.27	0.45*
1973	0.23	0.37	0.80*
1974	0.27	0.46	1.00*
1975	0.34	0.57	1.40
1976	0.48	1.42	1.60
1977	0.69	1.42	1.90

*estimated.
n.a.: not available.
Source: Robert Stobaugh and Daniel Yergin, eds., Energy Future (New York: Ballantine Books, 1979), p. 73. The Natural Gas Policy Act passed in 1978 consolidated inter- and intrastate markets, thereby changing pricing relationships.

and the gas was often used wastefully. Had natural gas prices never been regulated, natural gas would have been used more efficiently, gas would not have been withheld from interstate use, total production could have been higher, and the subsequent industrial gas curtailments would not have occurred.

Oil

For a better part of a century, public policy has been directed at keeping oil prices higher than would have occurred under competition. While oil demand rose steadily as new uses were found for this versatile fuel --from illumination and lubricants to locomotives, airplanes, cars, homes, and factories--the oil industry still faced periods of excess supply. Domestically, the Texas Railroad Commission and similar bodies in other states began to prorate to "conserve" oil resources, and to maintain higher prices. The Connally Hot Oil Act of 1935 was passed to enforce these conservation requirements by preventing interstate transport of oil in excess of the limits specified by the commission.[14]

By the 1950s, the price threat no longer came from excessive domestic production; instead, it came from cheap imports. The major oil multinationals had developed concession arrangements in Saudi Arabia, Kuwait, Iran, Iraq, and the United Arab Emirates. Domestic independents and oil state congressional delegations pushed hard for an oil import quota to dam the flood of cheap oil imports. Against his better judgment, President Eisenhower promulgated the mandatory oil import program in 1959, designed specifically to prevent large decreases in oil prices. In March 1966, the quota was lifted on residual fuel shipped to the East Coast, but continued for other products and crude oil until 1973. By then, riddled with exemptions from rising import demand, the quota was finally jettisoned. During its lifetime, the oil import quota drained domestic oil fields at a time when the United States could have taken advantage of cheap foreign oil.

By 1970, the world oil picture had changed completely. The United States no longer produced one-third of the world's oil; now it produced only about 20 percent. Even more serious, U.S. demand had grown so quickly that domestic production met only two-thirds of demand. Ironically, it was only one year after U.S. production had peaked that price controls were first slapped on as part of President Nixon's Economic Stabilization Program.

The role that oil price controls would play in U.S. energy policy was not perceived at the time. But after various extensions, price controls were codified into federal law through the 1973 Emergency Petroleum Allocation Act (EPAA) and extended until 1981 through the Energy Policy and Conservation Act of 1975 (EPCA).

Oil price controls had less impact on competitive prices than natural gas price controls. During the 1974-78 period, the average difference between controlled crude oil prices and the world oil price fell from 28 to 14 percent. The price for residual fuel oil, the oil product most competitive in the industrial market, was affected more by world residual fuel oil prices than by domestic price controls. Natural gas, on the other hand, cost only 30 percent as much as oil in 1973; not until 1982 did natural gas prices on many pipelines begin to approach those of residual fuel oil. Hence, natural gas price controls not only discriminated against coal use, but also provided a large price advantage over oil.

Coal

The federal government has not regulated coal prices, but it has regulated virtually every other aspect of coal use--from extraction and transportation to final combustion and disposal of waste products. The government controls workers' safety, the shape of the landscape after mining, the prices set by railroads for transporting coal (at least until passage of the Staggers deregulation legislation in 1980), the emissions allowed during coal burning, and even disposal of sludge and other wastes. Coal mining is one of the most dangerous and unhealthy occupations. Every stage of coal use holds the potential for damaging the environment, either by scarring the earth in strip mining or by polluting the air during combustion. The most compelling images of environmental pollution include the London "killer smog" in which thousands perished, strip mining in Appalachia, and today, acid rain in the Northeast. Government regulation did not arise from an over-zealous Congress or an aggressive bureaucracy, but rather from strong concerns across the country about real substantive problems.

From the beginning, it was the Clean Air Act which stirred the most controversy. Passed in 1970, the legislation established the framework for today's air pollution program, including the State Implementation Plan process, New Source Performance Standards for new facilities, statutory deadlines for compliance, and a 90 percent mandated reduction in automobile emissions. The 1977 amendments legislatively created protection for clean air areas, established more rigid standards for new facilities, and established a new set of deadlines. Despite such elaborate legislation, the Clean Air Act has not yet required stack gas scrubbers for industrial facilities, although such a requirement has been under review for some years. (Chapter 4 will discuss in more detail the effects of the Clean Air Act on coal use.)

Although federal regulations on safety and strip mine reclamation have affected mining productivity, the extent is subject to debate. It is true that the Coal Mine Health and Safety Act of 1969 decreased productivity in underground mines--but reductions in productivity were also caused by a depletion of the most cheaply producible surface mining reserves in the eastern United States and by labor problems. Because today's labor force is better trained and labor relations have stabilized, productivity is likely to increase in the future, even in the eastern coal fields.[15] Indeed, since 1975, real coal prices have declined.

When the Arab oil embargo hit in October 1973, the Senate began work on emergency legislation to convert power plants and large industrial facilities from oil to coal. That legislation, emerging as the Energy Supply and Environmental Coordination Act (ESECA), created the first mandatory coal conversion program. It authorized the Administrator of the Federal Energy Administration (precursor to the Department of Energy) to mandate the use of coal in utility and industrial boilers, provided that ambient air quality standards could be achieved. ESECA created a complex system of EPA approvals before prohibition orders were given, but exempted mandated facilities from new source performance standards, requiring only that their emissions comply with national ambient air quality standards.

Inevitably, ESECA failed. As noted earlier, coal prices rose faster in 1974 than price-controlled oil prices--thus eroding even further coal's competitive position. Any firm that found it economically unattractive to convert to coal in 1972 or 1973 found it even less attractive after enactment of ESECA. While the government became entangled in red tape, industry ordered new oil and gas burners at a rapid rate. Some firms on Federal Energy Administration conversion lists ultimately converted to coal, but it is unlikely that ESECA had any material impact on those decisions.

By 1977, the Carter Administration looked for stronger medicine to induce coal conversion. The major thrust of the Carter strategy was a proposed tax on industrial and utility oil and natural gas use. Ultimately, that tax would have raised industrial oil prices above the world price and industrial natural gas prices to that level. As a backup to the tax, the administration recommended a stronger regulatory program to force conversions. It failed to gain congressional enactment of the tax, but Congress did enact stronger regulatory authority--which became the Powerplant and Industrial Fuel Use Act (FUA).

That legislation made it difficult to build new oil and gas boilers and established a program for eliminating natural gas from utility boilers. In theory, installation of new oil and gas boilers above 100 million Btu per hour (mmBtu/hr) was precluded by federal law, although environmental and economic waivers were available. Utility gas use was theoretically prohibited by 1990. But gas use could be extended throughout most of the decade upon approval of a system compliance plan by the Department of Energy.

In a sense, the Fuel Use Act was obsolete the day it was passed. The natural gas shortages of the 1970-78 period were almost immediately transformed into a gas glut, as the Natural Gas Policy Act eliminated the economic restrictions that kept natural gas within the boundaries of producing states. By 1981, natural gas reserves slightly exceeded production. Demand was soft, as the natural gas market was weakened by conservation due to higher prices and economic recession. Congress subsequently eliminated the utility back-out requirements, and the Reagan Administration, contending that it was the marketplace that shaped industry decisions, promised extremely liberal implementation of the Fuel Use Act.

CONCLUSION

This historic retrospective and brief analysis of major federal energy regulatory activities was designed to help the reader understand the historic forces that drove fuel-use decisions and the role played by environmental standards in shaping those decisions prior to 1979. This history was traced to the 19th century in order to view the widest range of price, regulatory and technological changes. The specific focus on the 1973-81 period is particularly important, because the policies of that period heavily affect current industrial energy-use patterns.

A number of observations flow from this analysis. First, the switch away from coal use in the industrial sector has been occurring since World War II. Initially, coal was replaced by cheaper and more versatile oil. Later, as natural gas pipelines expanded across the country, gas cut into the markets of both coal and oil. During the mid-1960s, oil prices fell below coal prices in New England, resulting in a substantial switch from coal to oil. All of these major changes occurred before environmental standards were a serious concern.

Second, transportation costs were an important determinant of fuel use before 1973. Oil was the dominant industrial fuel in New England, coal was the most prominent source in the Mid-Atlantic states, and natural gas captured the market wherever pipelines were convenient to gas fields.

Third, boiler size and capacity utilization were critical factors in determining whether coal was feasible. Just as coal use declined in the industrial sector, utilities' use of coal increased dramatically. From 1951 to 1980, utilities' coal use expanded over 400 percent, causing an overall expansion in coal use of 40 percent.[16]

Fourth, while market forces generally explain the pre-1973 fuel mix, government policy shaped fuel-use choices after the Arab oil embargo. The imposition of price controls on crude oil resulted in a worsening of coal's competitive position with petroleum products. Even worse, the price controls imposed on interstate supplies of natural gas discouraged exploration, production and dedication of natural gas to that market, and caused a 1.8 tcf (50.4 billion cubic meters) loss of supplies to the industrial sector. Oil picked up about half that loss with lower demand accounting for the rest.

Fifth, the attempt to correct counterproductive government pricing policies with mandatory conversion authority did not work.

There appears to be little historical validity to the argument that environmental requirements prevented coal use. Coal use declined in the United States because other fuels were cheaper to use, more convenient, more flexible, and required less capital investment. Government price control policies in the 1970s encouraged further shifts away from coal. If no federal environmental legislation had been enacted, coal use in the industrial sector would have been only marginally affected.

This retrospective analysis has several limitations. It does not fully explain why the use of coal has not risen substantially since the oil price hikes of 1979-80 and the 1982 natural gas price increases, although low economic growth is clearly a major factor. It does not explain why coal was not used more in large industrial facilities, where economics would have been most advantageous. It does not offer the level of detail necessary for understanding specific, current or potential environmental problems, which, although not large in the aggregate, may nevertheless inhibit coal use in specific situations. It does not deal with the impact of prospective requirements for new industrial boilers and of many of the provisions of the 1977 amendments to the Clean Air Act. To delve more deeply into these issues, the next chapter will discuss current coal-use economics, with and without the requirements for stack gas scrubbers. From that discussion, we can then look at how actual fuel-use decisions are made and what role environmental quality has played in those specific decisions. Only then can we understand how environmental quality requirements have affected coal-use decisions across the country.

NOTES

1. Sam A. Schurr and Bruce C. Netschert, Energy in the American Economy, 1850-1975; An Economic Study of Its History and Prospects (Baltimore: Johns Hopkins Press, 1960), pp. 60-61.
2. Schurr, p. 36.
3. Schurr, p. 79.
4. The percentages are based on data in Department of Energy, Energy Information Administration, 1981 Annual Report to Congress, Vol. 2, May 1982, p. 9, Table 4 and p. 67, Table 30.
5. Schurr, p. 129.
6. EIA, 1981 Annual Report, p. 9, Table 4, and p. 107, Table 47.
7. EIA, p. 127.
8. U.S. Department of Energy, Energy Information Agency, Monthly Energy Review, Dec. 1982, p. 23.
9. In 1969, New England Electric System (NEES) converted its Brayton Point facility, the largest coal-fired plant in New England at that time, from coal to oil because of cheaper oil prices. NEES has recently converted this facility back to coal.
10. EIA, p. 67.
11. U.S. Department of Energy, Energy Information Administration, 1982 Annual Energy Review, April 1983, p. 67.
12. Craufurd D. Goodwin, ed., Energy Policy in Perspective: Today's Problems, Yesterday's Solutions (Washington, D.C.: Brookings Institution, 1981), p. 264.
13. Department of Energy, Energy Information Administration, 1980 Annual Report to Congress, Vol. 2, April 1981, p. 21.
14. Goodwin, p. 64.
15. Ronald M. Whitfield and Martin B. Zimmerman, "Productivity a Key to Coal's Future," Data Resources, Inc., Coal Review, Winter 1981-82, p. 37.
16. Department of Energy, Energy Information Agency, 1981 Annual Report to Congress, Vol. 2, May 1982, p. 127.

2
The Costs of Industrial Fuel Use

The economics of using coal in industrial boilers has changed dramatically since the substantial oil price increases resulting from the Iranian Revolution. Before then, coal was not economical in most industrial applications, except in cases where boilers were large, capacity factors were high, and coal transportation costs were low. The oil price increases improved the economics of using coal in all situations. In addition, the 1978 Energy Tax Act and the 1981 Economic Recovery Tax Act provided substantial new incentives for coal use. Recent reductions in real oil prices and increased costs for coal and coal transportation have narrowed these advantages, but coal use continues to have a substantial economic edge over oil.

It is easy to become confused about the costs of using different fuels. For example, in 1981, only 20 percent of the average cost of using coal was accounted for by the cost of coal at the minemouth. Coal is much more expensive to transport and requires larger, more expensive boilers, handling equipment, and pollution controls than oil or natural gas. In general, coal competes best in extremely large, continuous operations that can take advantage of economies of scale in both transportation and processing. Oil and gas, which are much more versatile and less capital-intensive than coal, can compete currently with coal in small and intermediate facilities. However, historically they had even developed strong positions in large boilers: Witness the 3.6 tcf (100.8 billion cubic meters) of natural gas and 1 mbd burned in utility boilers in 1981.[1]

The analysis in this chapter compares the costs of using oil and coal in industrial boilers. First it describes the technical limits to using coal, explaining in the process why this analysis focuses only on boilers. Next it compares the costs of using oil and coal in industrial boilers in 600 different situations. Such an analysis is designed to help the reader understand how sensitive coal-use economics are to such variables as plant size, capacity, transportation costs, and assumptions about future oil prices.

THE LIMITS TO COAL USE

For some applications, coal has inherent advantages. About half the coal in the industrial sector is used to provide the metallurgical feedstock for making coke, which, when heated with iron ore, becomes steel. Coal is the dominant fuel in the cement industry, where it is more efficient to use in kilns, and where the absorbency of the cement produced eliminates the need for emission controls. And coal dominates the electric utility industry, providing the energy source for over half of U.S. electric power generation.

But coal is not considered attractive for most other applications. In industrial process uses, for example, the heat generated during combustion can vary, and the particulates and other impurities can contaminate the product. Coal is used in just 5 percent of all process heaters, usually only in lime or cement kilns, and no substantial shifts in process use are foreseen. The data in Table 2.1, prepared by Energy and Environmental Analysis, Inc., for the Department of Energy, shows that unless there are substantial technical changes, only about .9 quad of coal (roughly equivalent to 450,000 barrels of oil per day) will be used in industrial process heaters by 1990. As Table 2.1 also shows, about 60 percent of all facilities expected to exist by then are already in place; at today's lower economic growth rates, even smaller future coal use is likely. Synthetic fuels can overcome technical barriers, but they are considerably more costly than oil and natural gas. Accordingly, use of synthetic fuels will not open up large new markets for coal unless oil prices skyrocket and stay there.

In theory, coal can be burned in all boilers, but cost restricts its use in small ones. Table 2.2 shows that in 1979, only 13 percent of boiler capacity was coal-fired in boilers under 100 mmBtu/hr and only 10 percent was coal-fired for boilers under 50 mmBtu/hr. Table 2.3 shows that boiler use dominates the food, textiles, and paper industries and flourishes in the chemicals and aluminum industries. In the industries with limited boiler use--steel and stone, clay and glass--coal already plays an important part either as a feedstock or as a fuel in process kilns. Because industrial boilers use over 10 percent of U.S. primary energy, they provide substantial potential for industrial coal use and hence are the focus of this chapter.

TABLE 2.1

PROJECTED COAL USE IN PROCESS HEATERS IN 1990
(10^{12} Btu)

	1990 Fuel Demand	
	Oil/Gas	Coal
Process Heaters:		
Built prior to 1981	3797	558
Built 1982-90	3080	369
TOTAL	6877	927

Source: EEA, IFCAM run generated December 17, 1979, published in U.S. Department of Energy, Energy Information Administration, Technical and Economic Feasibility of Alternative Fuel Use in Process Heaters and Small Boilers, February 1980, pp. 6-32.

TABLE 2.2

CAPACITY DISTRIBUTION OF SMALL INDUSTRIAL BOILERS BY SIZE AND FUEL TYPE
(mmBtu/hr)

Primary Fuel Type	<0.4	0.4-1.5	1.5-10	10-25	25-50	50-100	Total	%
Coal	4,100	14,260	25,250	26,280	75,980	95,200	241,070	13.0
Residual	10,300	42,520	117,840	98,660	154,120	121,650	545,090	29.5
Distillate	6,400	22,740	63,180	47,170	35,010	14,660	189,160	10.2
Natural Gas	26,400	88,830	199,740	164,700	210,810	182,880	873,360	47.2
TOTAL	47,200	168,350	406,010	336,810	475,920	414,390	1,848,680	99.9
Percent	2.6	9.1	22.0	18.2	25.7	22.4	100.0	

Source: PEDCO Environmental, Inc., "The Population and Characteristics of Industrial/Commercial Boilers," prepared for Environmental Protection Agency, May 1979, Tables 2-9, 2-11, 2-13, published in U.S. Department of Energy, Energy Information Administration, Technical and Economic Feasibility of Alternative Fuel Use in Process Heaters and Small Boilers, February 1980, Table 3.9, pp. 3-15.

TABLE 2.3

ENERGY CONSUMPTION BY INDUSTRY AND BY BOILER USE
(in quads)[1]
1974

Industry	Boiler	Total
Food	504.7	683.2
Textiles	149.7	186.8
Paper	923.1	1,136.9
Chemicals	1448.2	2,224.6
Petroleum Refining	366.1	1,448.5
Stone, Clay and Glass	6.4	1,055.8
Steel	366.0	1,116.6
Aluminum	261.0	476.2
Other[2]	1710.1	3,850.7
TOTAL	5735.3	12,179.3

[1]Excludes raw materials and feedstock uses. Excludes about 0.8 quads of wood residuals in the paper industry and about 0.3 quads of refinery gas in the petroleum refining industry.

[2]Includes miscellaneous manufacturing, agriculture, mining and construction industries.

Source: Energy Consumption Data Base, prepared by EEA for FEA, June 9, 1977, published in U.S. Department of Energy, Energy Information Agency, Technical and Economic Feasibility of Alternative Fuel Use in Process Heaters and Small Boilers, February 1980, Table 3.3, pp. 3-5 and Table 3.6, pp. 3-10.

ELEMENTS FOR COAL USE

Capital, transportation, and operating costs dominate the end-use costs of using coal, but are not important for oil. Table 2.4 shows that the price of residual fuel oil is over 80 percent of the total cost of using it. The minemouth price of coal, however, is only 31 to 27 percent of the total coal-use cost. Even when transportation costs are added, delivered coal prices equal no more than 43 to 50 percent of the total cost. It is the capital and operating costs that add so much to the cost of using coal, representing 50 to 55 percent of the total for plants not requiring scrubbers and even more if scrubbers are required.

The use of different assumptions from those in Table 2.4 could lead to dramatically different conclusions. For example, if one were to assume smaller boiler capacities, longer transportation hauls, and lower oil prices, oil would be cheaper to use than coal. Before showing how these different assumptions change relative to economics, let us examine each component of the cost of using coal.

Delivered coal prices vary substantially, based on region, coal quality, and transportation cost. Table 2.5 shows almost a fourfold difference between the lowest and highest minemouth prices for low sulfur coal, ranging from $11.20 per ton (.907 metric tons) in Montana-Wyoming to $40.64 per ton (.907 metric tons) in Northern Appalachia. Some of these differences stem from quality characteristics and others from distance to market. Western coal is cheap because it has a low energy content and is far from most markets. These characteristics offset the somewhat higher price the fuel could command because of its low sulfur content.

Varying transportation costs can cause a huge difference in coal prices. Table 2.6, which sets forth transportation costs between regions, shows costs as low as $9 per ton (.907 metric tons) <u>within</u> the Rocky Mountain states, and as high as $20 per ton (.907 metric tons) <u>between</u> the Rocky Mountain states and the Midwest or between Southern Appalachia and New England. Taking into account both minemouth and transportation costs, an electric utility in the West might pay less than $20 per ton (.907 metric tons) for coal while a New England firm may pay over $50 per ton (.907 metric tons).

Coal operation and maintenance (O&M) costs (including staff, waste disposal, power and miscellaneous expenses) also contribute to total cost. In the example shown, they represent 12 to 14 percent of total cost--and that amount increases substantially when scrubbers are required.

Table 2.4 shows that capital costs make up from 38 to 43 percent of coal-use costs, but only about 13

TABLE 2.4

COMPARISON OF COAL AND OIL COSTS
(1980 dollars per million Btu)

	Coal Boiler		Oil-Natural Gas Boiler
	Midwest Site	New England Site	
Fuel			
Minemouth	1.38	1.38	-
Transportation	.56	1.15	-
Fuel Oil	-	-	5.47
Operation and Maintenance			
Base	.61	.61	.36
For Scrubbers	.35	.35	-
Capital			
Base	1.93	1.93	.84
For Scrubbers	.83	.83	-
Total	4.48	5.07	6.67
Total (with Scrubbers)	5.66	6.25	n.a.

Assumptions: S. Appalachian low sulfur coal; .70 capacity factor; 170 mmBtu/hr. boiler; mid-range fuel prices. Expenses are annualized per million Btu per input fuel over a 30-year period, which is the estimated lifetime of the plant.

n.a. = not available.

Source: Data Resources, Inc., provided fuel prices. ICF, Inc., supplied capital and operation-and-maintenance costs.

TABLE 2.5

MARGINAL MINEMOUTH COAL PRICES
(1980 dollars per ton)

Supply Region and Sulfur Type	1979	1980	1981	1982
North Appalachian				
Low Sulfur	24.48	31.23	35.46	40.64
Medium Sulfur	25.61	29.12	32.56	35.84
High Sulfur	20.12	22.77	25.27	27.71
South Appalachian				
Low Sulfur	28.46	31.80	35.65	40.15
Medium Sulfur	26.33	29.95	33.20	36.72
High Sulfur	20.58	23.28	25.76	28.75
Midwest				
Low Sulfur	28.35	31.35	34.60	39.05
Medium Sulfur	26.32	28.70	31.30	34.56
High Sulfur	23.46	25.75	27.70	29.93
Montana-Wyoming				
Low Sulfur	8.74	9.58	10.27	11.20
Medium Sulfur	8.68	9.52	10.20	11.11
High Sulfur	8.68	9.52	10.20	11.11
Colorado-Utah				
Low Sulfur	18.15	20.43	21.36	23.51
Medium Sulfur	17.21	19.27	19.92	22.14
High Sulfur	17.21	19.27	19.92	21.65

Source: Data Resources, Inc., Coal Review Update, Spring 1982, p. 25.

TABLE 2.6

COAL TRANSPORTATION COSTS
(1980 dollars per ton)

Destination	Mine Location	
	South Appalachia	Montana/Wyoming
New England	$20.68	-
Mid-Atlantic	15.82	-
Midwest	8.66	$19.45
Central Rocky Mountain	-	9.41
Pacific	-	24.87

Source: Data Resources, Inc., Coal Review Update, Spring 1982, p. 18.

percent of oil use costs. Coal boilers without scrubbers typically cost more than three times as much as oil boilers without scrubbers for facilities of this size; coal boilers with scrubbers can cost 4.4 times as much as oil boilers with scrubbers. Coal facilities exhibit economies of scale since some equipment is necessary for large and small facilities alike. As shown in Table 2.7, the 370 mmBtu/hr boiler has approximately a 30 percent unit cost advantage over the smallest facilities.

Because of coal's high capital costs, its use is extremely sensitive to capacity utilization. If a coal-fired boiler is used intermittently, fixed capital costs must be spread over fewer units of useful energy, and unit costs will rise. Conversely, if the facility is used most of the time, unit costs will fall.

So far, we have addressed only the generic differences that affect the cost of using coal. Individual boiler costs also vary substantially, even when capacity is identical. One study found that cost estimates for similar-sized boilers differed as much as 37 percent for coal boilers and 46 percent for oil boilers.[2] Average costs can also be affected by site-specific factors such as lack of space or extra foundation work.

Extraordinary variations in end-use costs do not occur in the case of oil. Since fuel costs represent most of the total expense in using oil, and oil fuel costs are largely uniform across the United States, the unit cost of firing a small oil-fired boiler in New England does not differ substantially from that of heating a large oil-fired boiler in Wyoming. (As gas prices reach residual fuel oil equivalent levels, natural gas prices will also be more consistent among most regions.) However, the same New England facility might face end-use costs 50 to 75 percent higher than a Wyoming facility if they both burned coal. Hence, location of facilities can be another important factor in a decision to turn to coal.

THE ECONOMIC COMPARISON

In order to compare coal's relative costs over this entire range of parameters, we constructed a computer-assisted, discounted cash-flow model. The model was developed to calculate 600 different cost comparisons of coal and oil. Each of the resulting tables compares two coal sources, three oil price scenarios, three boiler sizes, and five capacity limits. They show how fuel and transportation costs, O&M costs, capital costs, and capacity utilization when combined, determine whether coal has a competitive advantage or disadvantage over oil.

TABLE 2.7

ECONOMIES OF SCALE
(thousands of 1980 dollars)

	Capital	Capital Costs per mmBtu
Dual Fuel Boilers (Oil and Gas)		
70 mmBtu/hr	$1,740	$25
170 mmBtu/hr	3,050	18
370 mmBtu/hr	5,285	14
Ratio of 370 to 70:	3.0	-
Coal		
70 mmBtu/hr	$4,500	$64
170 mmBtu/hr	8,490	49
370 mmBtu/hr	16,600	45
Ratio of 370 to 70:	3.7	-

Source: ICF, Inc., supplied figures for capital costs.

Three different oil price trajectories were used for this analysis. The low-price scenario envisages declining real oil prices until 1990. This scenario--once considered a virtual impossibility--looks more plausible today. The mid-price scenario assumes that oil prices will level in real terms but then rise by the end of the decade. The high-price scenario envisages a significant oil supply interruption in 1985, which drives prices rapidly up to $45 per barrel; prices fall back to $42 per barrel by the end of the decade, but then continue to rise over the next decade. A few years ago, this scenario seemed plausible, but currently seems very unlikely.

The analysis also varied the distance from the coal source to demonstrate how coal transportation costs affect total costs. Industrial sites were assumed to be either near the mine (e.g., a southern facility using Southern Appalachian coal) or at a considerable distance (e.g., a midwestern facility using Rocky Mountain coal). Because railroads might attempt to capture some of the rents from larger differences between coal and oil prices, the transportation costs are indexed to oil prices.

Coal economics are examined using oil prices characteristic of the period before the Iranian Revolution, when crude oil prices were considerably less than they are today. They are then examined using early 1982 oil prices. Table 2.8 reflects the earlier period by assuming fuel prices estimated in the 1978 EIA Administrator's Annual Report and eligibility for only a 10 percent investment tax credit (ITC). (The Energy Tax Act of 1978 provided an additional 10 percent ITC for coal-fired boiler costs and removed the 10 percent ITC for oil-fired equipment.) Table 2.9 reflects the base case assumptions discussed in this chapter, including the more generous depreciation provisions of the Economic Recovery Tax Act of 1981. Finally, the analysis focused on the effect of an across-the-board requirement for stack gas scrubbers using the base case assumptions. Table 2.10 shows coal's attractiveness compared to oil when scrubbers are required on all coal-fired industrial boilers, no matter what the size.

Costs for different-sized plants were fed into the computer. The analysis assumed that package boilers would be used for smaller plants, but that they would be tailor-made for larger facilities. Different construction periods were assumed, with small package oil boilers requiring only a year for construction and larger, field-erected coal boilers requiring two to three years. Three different boiler sizes were used: 70 mmBtu/hr, 170 mmBtu/hr, and 370 mmBtu/hr. Finally, we compared five different capacity utilization levels to determine costs for each of the boiler sizes.

TABLE 2.8
RESULTS OF THE ECONOMIC ANALYSIS FOR 1978 FUEL COSTS

Oil Price / Plant Location / Type of Coal Burned / Capacity Factor / Plant Size		LOW				MID				HIGH			
		Distant		Local		Distant		Local		Distant		Local	
		S.A.	R.M.	S.A.	R.M.	S.A.	R.M.	S.A.	R.M.	S.A.	R.M.	S.A.	R.M.
70 mmBtu per* hour	.25	-2.48	-1.95	-2.44	-1.46	-2.02	-1.44	-2.04	-0.97	-1.31	-0.58	-1.49	-0.17
	.4	-1.14	-0.61	-1.04	-0.06	-0.69	-0.10	-0.65	0.42	0.02	0.75	-0.10	1.22
	.55	-0.54	-0.00	-0.41	0.57	-0.08	0.51	-0.02	1.06	0.63	1.36	0.53	1.86
	.7	-0.19	0.35	-0.05	0.93	0.27	0.85	0.34	1.42	0.98	1.71	0.89	2.22
	.85	0.04	0.57	0.19	1.16	0.49	1.08	0.58	1.65	1.20	1.93	1.13	2.45
70 mmBtu per hour	.25	-3.67	-3.13	-3.65	-2.67	-3.21	-2.62	-3.26	-2.18	-2.50	-1.77	-2.70	-1.38
	.4	-1.89	-1.35	-1.80	-0.82	-1.43	-0.84	-1.41	-0.33	-0.72	0.01	-0.86	0.47
	.55	-1.08	-0.54	-0.96	0.02	-0.62	-0.03	-0.57	0.51	0.09	0.82	-0.02	1.31
	.7	-0.61	-0.08	-0.48	0.50	-0.16	0.43	-0.09	0.99	0.55	1.29	0.46	1.79
	.85	-0.31	0.22	-0.17	0.81	0.14	0.73	0.22	1.30	0.85	1.53	0.77	2.10
170 mmBtu per hour	.25	-3.20	-2.66	-3.20	-2.23	-2.74	-2.16	-2.81	-1.74	-2.03	-1.30	-2.26	-0.94
	.4	-1.59	-1.05	-1.52	-0.54	-1.13	-0.54	-1.13	-0.05	-0.42	0.31	-0.58	0.75
	.55	-0.85	-0.32	-0.75	0.23	-0.40	0.19	-0.36	0.72	0.32	1.05	0.19	1.52
	.7	-0.43	0.10	-0.31	0.66	0.02	0.61	0.08	1.15	0.73	1.47	0.63	1.95
	.85	-0.16	0.37	-0.03	0.95	0.29	0.88	0.36	1.44	1.01	1.74	0.91	2.24
370 mmBtu per hour	.25	-2.93	-2.40	-2.93	-1.95	-2.48	-1.89	-2.53	-1.46	-1.76	-1.03	-1.98	-0.66
	.4	-1.42	-0.89	-1.35	-0.37	-0.97	-0.38	-0.96	0.12	-0.26	0.48	-0.41	0.92
	.55	-0.74	-0.20	-0.63	0.35	-0.28	0.31	-0.24	0.83	0.43	1.16	0.31	1.63
	.7	-0.35	0.19	-0.22	0.76	0.11	0.70	0.17	1.24	0.82	1.55	0.72	2.04
	.85	-0.09	0.44	0.04	1.02	0.36	0.95	0.43	1.51	1.07	1.81	0.98	2.31

All figures represent the difference in annualized cost per mmBtu of fuel input between a coal-fired and an oil-fired boiler in 1980 dollars. Minus numbers indicate a negative cost differential for coal-fired facilities.

* Assumes a 20 percent investment tax credit (ITC) for the coal equipment, and no ITC for the oil-fired equipment. All other figures assume a 10 percent ITC for all boilers.

TABLE 2.9
RESULTS OF THE ECONOMIC ANALYSIS, USING BASE-CASE ASSUMPTIONS

Oil Price / Plant Location / Type of Coal Burned / Capacity Factor / Plant Size		LOW				MID				HIGH			
		Distant		Local		Distant		Local		Distant		Local	
		S.A.	R.M.	S.A.	R.M.	S.A.	R.M.	S.A.	R.M.	S.A.	R.M.	S.A.	R.M.
	.25	-1.22	-0.69	-1.18	-0.20	-0.79	-0.20	-0.80	0.27	1.01	1.74	0.83	2.16
70	.4	0.11	0.65	0.21	1.19	0.55	1.14	0.59	1.66	2.35	3.08	2.23	3.55
mmBtu	.55	0.72	1.25	0.85	1.82	1.16	1.75	1.22	2.30	2.96	3.69	2.86	4.18
per	.7	1.07	1.60	1.21	2.19	1.51	2.09	1.58	2.66	3.30	4.04	3.22	4.54
hour	.85	1.29	1.83	1.44	2.42	1.73	2.32	1.82	2.89	3.53	4.26	3.45	4.78
	.25	-0.99	-0.45	-0.96	0.01	-0.55	0.04	-0.59	0.49	1.25	1.98	1.05	2.37
170	.4	0.27	0.80	0.35	1.33	0.71	1.30	0.73	1.80	2.51	3.24	2.37	3.69
mmBtu	.55	0.84	1.37	0.95	1.93	1.28	1.87	1.33	2.40	3.08	3.81	2.97	4.29
per	.7	1.17	1.70	1.30	2.27	1.60	2.19	1.67	2.74	3.40	4.13	3.31	4.63
hour	.85	1.38	1.91	1.52	2.49	1.82	2.40	1.89	2.97	3.61	4.34	3.53	4.85
	.25	-0.82	-0.29	-0.79	0.19	-0.38	0.20	-0.42	0.66	1.41	2.14	1.22	2.54
370	.4	0.36	0.90	0.46	1.43	0.80	1.39	0.83	1.91	2.60	3.33	2.47	3.79
mmBtu	.55	0.90	1.44	1.02	2.00	1.34	1.93	1.40	2.47	3.14	3.87	3.04	4.36
per	.7	1.21	1.75	1.35	2.32	1.65	2.24	1.72	2.80	3.45	4.18	3.36	4.68
hour	.85	1.41	1.95	1.56	2.53	1.85	2.44	1.93	3.01	3.65	4.38	3.57	4.89

All figures are the difference in the annualized cost of boiler operation, per mmBtu of fuel input in 1980 dollars.

TABLE 2.10
RESULTS OF THE ECONOMIC ANALYSIS, USING BASE-CASE ASSUMPTIONS, AND A REQUIREMENT THAT SULFUR DIOXIDE CONTROLS (SCRUBBERS) BE EMPLOYED

Plant Size	Capacity Factor	LOW				MID				HIGH			
Oil Price		Distant		Local		Distant		Local		Distant		Local	
Plant Location / Type of Coal Burned		S.A.	R.M.	S.A.	R.M.	S.A.	R.M.	S.A.	R.M.	S.A.	R.M.	S.A.	R.M.
70 mmBtu per hour	.25	-4.53	-4.00	-4.39	-3.41	-4.09	-3.50	-4.01	-2.94	-2.29	-1.56	-2.38	-1.05
	.4	-2.05	-1.52	-1.87	-0.90	-1.61	-1.02	-1.50	-0.43	0.19	0.92	0.14	1.46
	.55	-0.92	-0.39	-0.73	0.24	-0.48	0.11	-0.36	0.72	1.32	2.05	1.28	2.60
	.7	-0.28	0.26	-0.08	0.90	0.16	0.75	0.30	1.37	1.96	2.69	1.93	3.26
	.85	0.14	0.67	0.34	1.32	0.58	1.17	0.72	1.79	2.38	3.11	2.36	3.68
170 mmBtu per hour	.25	-3.91	-3.37	-3.80	-2.82	-3.47	-2.88	-3.43	-2.35	-1.67	-0.94	-1.79	-0.47
	.4	-1.64	-1.11	-1.50	-0.52	-1.20	-0.61	-1.13	-0.05	0.60	1.33	0.51	1.83
	.55	-0.61	-0.08	-0.46	0.52	-0.17	0.42	-0.08	0.99	1.63	2.36	1.56	2.88
	.7	-0.02	0.51	0.14	1.12	0.42	0.00	0.52	1.59	2.21	2.95	2.15	3.48
	.85	0.36	0.89	0.53	1.51	0.80	1.39	0.90	1.98	2.60	3.33	2.54	3.86
370 mmBtu per hour	.25	-3.44	-2.90	-3.31	-2.34	-3.00	-2.41	-2.94	-1.86	-1.20	-0.47	-1.30	0.02
	.4	-1.36	-0.82	-1.20	-0.23	-0.92	-0.33	-0.83	0.25	0.88	1.61	0.81	2.13
	.55	-0.42	0.12	-0.24	0.73	0.02	0.61	0.13	1.21	1.82	2.55	1.77	3.09
	.7	0.12	0.66	0.30	1.28	0.56	1.15	0.68	1.75	2.36	3.09	2.32	3.64
	.85	0.47	1.01	0.66	1.64	0.91	1.50	1.03	2.11	2.71	3.44	2.67	4.00

All figures represent the difference in annualized costs per mmBtu of fuel input between a coal-fired and an oil-fired boiler in 1980 dollars.

The discount rates play an important part in this type of analysis. For the purposes of this evaluation, a nominal discount rate of 15.4 percent was used--a relatively conservative assumption. This rate corresponds to an after-tax rate of return on equity of 19 percent (assuming a 70:30 equity-to-debt ratio for a firm), and an even higher rate before taxes, depending on the firm's marginal tax rate. There are two reasons such a high discount rate was used. First, private firms tend to use high rates for projects of this type, often expecting a payoff in one or two years. Second, the use of a high discount rate partially compensates for industry's reluctance to convert to coal.

ANALYTICAL RESULTS

The analytical results are consistent with common sense. Under most conditions, coal was not very attractive in 1978--before the 1979-80 oil price hikes, crude oil decontrol, the added 10 percent investment tax credit for coal conversion, elimination of that credit for oil- and gas-fired facilities, and the five-year depreciation allowed under the Economic Recovery Tax Act of 1981. Coal would, of course, have been much more promising if industrial managers had been able to foresee the large price hikes resulting from the Iranian Revolution. By 1981, the economics of coal had improved dramatically. The accelerated depreciation provisions in the Energy Tax Act lowered the cost of all capital, including capital for coal conversion. But the biggest change was the more than 170 percent oil price hike resulting from the Iranian interruption.

Tables 2.8, 2.9, and 2.10 illustrate vividly how coal and oil compare on the basis of the parameters discussed earlier. Table 2.8 shows that prior to 1979, coal had strong economic advantages only when capacity levels were high and distance from the coal fields was small. The greater incidence of positive numbers in Table 2.9, the base case, indicates that since the 1979-80 oil price hike, coal's relative economic position has improved markedly. Coal was unattractive only when capacity levels were extremely low.

Of the many parameters considered in the base case, the two most important in comparing coal and oil use were the level of world oil prices and the capacity of the facility. For example, at low world oil prices, the cost of using Appalachian coal in a 70 mmBtu/hr boiler (at 55 percent capacity) would be 72 cents per mmBtu _less_ than the cost of using oil in the same-sized boiler. At high world oil prices, coal would cost $2.96 per mmBtu less. Because the low- and mid-price scenarios for the 1990s are similar, the economics do not differ dramatically. But considering that much of the East Coast and even parts of the Midwest would be expected to import substantial amounts of Appalachian coal, even these small differences could affect decisions. At the high world oil price levels, coal's economic advantages are enormous.

Capacity differences affect coal economics even more dramatically. For example, if the same 70 mmBtu/hr boiler were used at 25 percent capacity under the mid-price oil scenario, oil would be cheaper to use than coal by 79 cents per mmBtu. However, if that boiler were used at 85 percent capacity, coal would be cheaper by $1.73 per mmBtu (a difference of $2.52). This analysis demonstrates that coal is mainly used in facilities operated at high capacity levels and the common sense conclusion that unit costs rise rapidly as capacity decreases.

Other economic factors, such as access to western coal, transportation costs, and economies of scale, are important but not nearly as decisive. In general, the cost difference between distant and local sites was about 70 cents per mmBtu, and access to western coal resulted in a difference of 60 cents per mmBtu.

The economies of scale were surprisingly small, relative to other factors--but they were more pronounced at lower capacities. Considering the historic correlation between boiler use and size, it is surprising at first that economies of scale have so little importance here. However, when one considers that the huge tax advantages available for using coal effectively result in eliminating half of the capital costs for the firm, the diminished role of scale economies becomes clear.

Table 2.10 illustrates the impact of a stack gas scrubber requirement on all industrial facilities. Such a requirement would have a measurable impact on coal use in the low- and mid-price scenarios, but virtually no adverse effect in the high-cost scenario. If such a prospective EPA scrubber requirement were tied to larger boilers, then its effect could be reduced substantially. Indeed, if the smaller boilers were exempt and all firms made decisions based exclusively on costs, then the scrubber requirement need not have any appreciable effect on coal use based purely on economics.

CONCLUSION

It is clear that coal economics have improved dramatically since the more than doubling of oil costs, the decontrol of crude oil prices, and the implementation of tax advantages provided by the Energy Tax Act of 1978 and the Economic Recovery Tax Act of 1981.

Since natural gas prices will soon rise to levels comparable to the costs of residual fuel oil, a close competitor in the boiler market, this analysis gives coal a long-term competitive advantage over natural gas as well. The one public policy step that could diminish coal's relative position--but certainly not push it back to the pre-Iranian Revolution era--is a stack gas scrubber requirement on all new coal boilers. But even such a requirement need not cripple coal use.

By comparing costs of building new oil and coal boilers, the analysis does not directly compare the economics of coal conversion. Only in cases where the oil boiler is essentially obsolete will such a comparison be fully applicable. Nevertheless, since oil costs represent the overwhelming costs of using that fuel and since most conversion decisions will involve oil boilers with a substantial amount of their useful life already depreciated, the analysis represents a useful approximation of conversion economics for many cases.

Considering the economic advantages of using coal, one would expect a much greater shift toward coal use in the industrial sector. High interest rates, sluggish economic activity, low natural gas prices, and lack of business confidence are some of the reasons why coal conversion is not taking place today. But one would still expect much greater activity in strategic planning for coal conversion and a greater volume of feasibility studies and engineering work than is taking place. Clearly, some very cost-effective decisions are not being made--and some basic reasons for this apparent lack of enthusiasm can be traced to the decision-making process itself. That process is the subject of the next chapter.

NOTES

1. U.S. Department of Energy, Energy Information Administration, Monthly Energy Review, December 1982, p. 68.

2. U.S. Department of Energy, Industrial Boiler Costs: A Comparison of Recent Studies, DOE/EIA-0183/12, September 1979.

3 The Process of Industrial Fuel-Use Decisions

Real world industrial decisions on which fuels to use vary substantially from the predictions of economic modelers. The reason is simple. Modelers assume that industry will always choose the cheapest fuel, but will be indifferent to the mix of capital and management commitment associated with the alternatives. In the real world, potential capital investments in energy facilities must compete with all other corporate priorities--not with some modeler's assumption. Coal conversion projects, for example, must be compared with projects to modernize productive facilities, expand production, and introduce new projects. Expenditures for purchasing oil and gas may be approved at low levels in a corporation; but a coal conversion project, because it involves such large capital expenditures, would normally be approved by a firm's board of directors. These factors create a bias against coal use. Only when energy costs become a serious competitive factor in an industry--the cement industry being the only good current example--will coal conversion projects receive high priority.

That decisions about capital expenditures are made differently from decisions about routine operating expenses is well documented in the literature, but these considerations are not built into econometric projections of energy demand by fuel source.[1] In a sense, this failure is understandable. There is no good way to quantify the opportunity costs for a myriad of individual firms to understand what discount rates and other considerations would be applied to coal conversion investments against other capital investments.

This chapter is designed to clarify the industrial decisionmaking process as it affects coal conversion. By examining in detail both how the process works and how individual firms go about fuel-use decisions, we intend to point out the importance of various factors in shaping industrial fuel-use choices, including the impact of environmental requirements.

INDUSTRIAL CAPITAL DECISIONS: THE TORTUOUS PATH

The industrial capital budgeting process sets strategic objectives for allocating resources, and for gaining a certain amount of management consensus on objectives and future accomplishments. The process requires that individual capital proposals be reviewed at many levels in a corporation: At the plant, by staff and line officials, by management committees, and ultimately by the board of directors. As projects move up and around this process, consensus and coalition-building become critical elements in developing ultimate claims on the capital budget. Except for obligatory projects, such as meeting mandated environmental requirements, the capital allocation process is extremely competitive.

Capital budgeting has a tremendous impact on industrial fuel-use decisions. Most energy expenditures do not traverse the capital budgeting process; rather they can be approved by operating officials. Plant or division management can tighten up energy housekeeping, make conservation investments, switch between gas and oil, and take other steps to control energy costs without top management control. Because of the capital-intensive nature of coal conversion investments, however, they are almost inevitably considered in the competitive, capital budgeting process. A different set of institutional actors becomes involved in these projects, moving decisions further away from simple technical and economic factors. This difference in the decisionmaking process, as we shall see, profoundly affects the substance of energy conversion decisions.

To help the reader understand the process, it has been broken down into four major components (as set forth in Table 3.1). Initiating the review is the first step, requiring some corporate official or officials to determine that a potential problem exists in current fuel-use patterns. Most potential coal conversions never even reach this stage. Determining technical feasibility represents the second phase. Selecting contestants for the capital budgeting process represents the third component. This straightforward component consists of analyzing various alternatives to determine the economic value of the proposed coal conversion compared to the current fuel-use patterns. At this stage, management must decide whether to recommend a capital allocation for coal conversion against other priorities. Making the final decision concludes the process. At this stage, the highest levels of management must decide how an expensive

TABLE 3.1

A PROCESS MODEL OF INDUSTRIAL FUEL-USE DECISIONS

Initiating Review	Determining Feasibility	Selecting Contestants	Biting the Bullet
1. Competitive nature of the industry	1. Site-specific capital costs	1. Economic viability	1. Availability of discretionary capital
a. Energy costs as a % of total costs	2. Space limitations	a. Discounted cash flow	2. Payoff of coal conversion project compared to others
b. Intermediate vs. end-use	3. Process characteristics	b. Payback period	3. Management attitudes/policy
c. Actions of competitors	4. Economies of scale	c. Rate of return	4. Competitive nature of the industry
d. Corporate strategic goals	5. Feasibility of unconventional alternatives	2. External stimuli	a. Actions of competitors
2. Character of firm		a. Environmental requirements	b. Intermediate vs. end-use
a. Management attitudes/policy		b. Expiring natural gas contracts	c. Energy costs as a % of total costs
b. Experience of energy staff		c. Fear of supply insecurity/price spike	5. Uncertainties
c. Corporate organization for energy		d. Expansion or plant overhaul	a. Questions about future cost relationships
3. External stimuli		3. Corporate policy	b. Government policies
a. Environmental requirements		4. Competing projects under purview of decisionmaker	c. Environmental delays
b. Expiring natural gas contracts			d. Security of coal supplies
c. Fear of supply insecurity/price spike			6. External stimuli
d. Expansion or plant overhaul			a. Environmental requirements
			b. Expiring natural gas contracts
			c. Fear of supply insecurity/price spike
			d. Expansion or plant overhaul

coal conversion project compares with other corporate and fiscal objectives of the company.

At each stage, a different set of decisionmakers and staff will be involved with a potential project, which includes both proponents and opponents. Our interest is not in describing the process in any detail, but in understanding at each stage what considerations are most important in making industrial fuel-use decisions. The bulk of this chapter will focus on those considerations.

BEGINNING THE PROCESS: THE FIRST STEP MAY BE THE HARDEST

The failure of many firms to perceive that cheaper and more secure energy options are available to them may be the single biggest barrier to coal use. Few firms have converted to coal since the 1973 Arab oil embargo, and most potential conversions have never even been seriously reviewed. Although the economic incentives to switch to coal prior to 1979 were not overwhelming, they were certainly stronger than the actual number of conversions. When firms evaluated coal conversion possibilities, they found that they were economic. In the event of higher oil prices, industrial firms should review their facilities and convert many oil-fired boilers to coal.

Clearly, a firm's competitive position is one, if not the major, determinant of its interest in switching to coal.[2] Firms that compete in consumer goods end-use markets--such as food and housewares--are primarily concerned about maintaining or expanding their market share in small segmented oligopolistic markets, such as the corn flake and toothpaste markets. Competition revolves more around brand name and product identification than on prices. To retain market share in such industries, the development and marketing of new products may be considerably more important than holding down costs. On the other hand, producers of homogeneous commodities such as aluminum and cement compete more on the basis of cost. Firms in these industries cannot afford substantially higher energy costs than their competitors without suffering loss of profits and market share.

While most industries' fuel costs are less than 7 percent of total costs, certain industries such as cement and aluminum face fuel costs up to 33 percent of the value of their products. Steel's energy costs are 22 percent of its total costs. Paper, petroleum, and chemicals face energy costs from 7 to 10 percent of the value of shipments, a smaller but not unsubstantial portion of total costs. Aggregate figures can be misleading for some industries, however. Textiles show relatively low energy costs as a percentage of product costs, but in certain segments of the industry, such as carpet manufacturing, energy use and costs are high. Likewise, certain parts of chemical and pulp and paper manufacturing experience high energy costs.

The ratio of energy costs to total product costs is not the only measure for determining whether energy costs pose competitive pressures. If all firms in a particular industry fail to shift fuels, even when energy costs are high, the direct competitive effect is minimized. All firms would have higher costs, and market shares would be preserved. In theory, less of the commodity would be demanded, greater substitution would occur, and total profits would drop, but all firms would suffer equally. It is when some firms shift fuels and others do not that energy costs may pose competitive pressures.

The evidence for this competitive hypothesis is mixed. In the cement industry, where energy costs represent a large portion of total costs, coal use has risen rapidly from about 35 percent of total fuel consumption in the early 1970s to about 80 percent today. While Ideal Cement--the company participating in this study--initially undertook conversions to improve the reliability of its natural gas supplies, the economics were so favorable that it decided to make a long-term commitment to coal. As Ideal surged ahead, other firms in the industry followed suit briskly. The outcome in the brick industry has been dramatically different, despite similar processes. Although General Shale converted all of its brick plants to coal, the industry as a whole has been very slow to make changes. Because of high transportation costs, the brick industry is extremely regional in character, which may partly explain why fuel costs appear to be such a weak competitive factor. However, the cement industry has similar characteristics but has made the move to coal.

Some firms do face competitive pressures to convert, although these pressures are either not terribly strong or are unique in character. A proposed conversion from oil to a mixture of wood and coal by St. Regis at its Bucksport, Maine, mill reflected some pressures in the pulp and paper industry to reduce oil use. Likewise, several conservation and coal conversion projects previously slated by Kaiser Aluminum for its Gramercy, Louisiana, facility reflected an assessment of what would be necessary for the firm to compete successfully in the aluminum business once natural gas prices reached market levels. Hence, competitive forces within an industry--the importance of price competition compared to other factors, the percentage of energy costs to total product costs, and the energy-related actions of competitors--will be a stimulus for some firms to review fuel use at existing facilities.

But for most firms, competition will not act as a strong stimulus.

The intensity of these competitive concerns may be reflected in a firm's internal organization and corporate energy policies. Almost all large corporations have instituted energy committees, but their use varies substantially.[3] In some cases, the energy committee does little more than review the firm's conservation progress. On the other hand, Du Pont maintains an energy committee that sets energy policy and reviews firm-wide energy use. As the major corporate energy advisory body, this committee has been instrumental in promoting Du Pont's corporate energy goals. An effective energy committee may be supplemented by internal corporate energy staff. In some cases, competent and aggressive energy staffs actually initiate policies and projects rather than merely respond to line management.

The most dramatic and clear-cut corporate energy commitment occurs in the setting of quantitative strategic goals. Some firms have established deadlines to replace all use of oil and natural gas uses. This strategic goal-setting establishes in advance the rough outline for energy review of individual plants. Du Pont set a goal to convert 95 percent of its coal-capable plants, now burning oil or natural gas, back to coal. General Shale decided to convert all of its plants to coal by 1981 and Ideal Cement started in the early 1970s to convert every possible facility. General Foods' management voiced a strategy of installing dual energy capabilities (oil/gas) in all plants, while St. Regis' corporate policy is to "back-out" fuel oil consumption to the extent possible.

Corporations with a strong commitment to coal use have usually had fairly extensive experience with coal plants. Du Pont, for example, actively pushed coal conversion after the 1973 Arab oil embargo. Ideal Cement began at the same time, when a former federal energy czar took over as chief executive officer of the company. As these corporations gained experience, they developed corporate goals and actively sought out conversion opportunities.

In most cases, the decision to review fuel use at a particular facility was the result of external stimuli. Kaiser Aluminum, for example, initiated a review of its Louisiana plants in the mid-1970s because of concern about the costs and reliability of natural gas supplies. It accelerated this review when the Carter Administration proposed a tax on oil and natural gas use and continued it after legislation to decontrol new natural gas prices passed through Congress. The imminent expiration of several natural gas contracts further prompted Kaiser to begin engineering for conversion to coal at its Gramercy facility, although the conversion was never completed due to the recession and the increase in natural gas supplies. General Shale and Ideal Cement experienced several gas curtailments in the early 1970s that stimulated their initial interest in coal. Exxon, at its Baton Rouge refinery, became so concerned about its external steam supply from Gulf State Utilities, that it began to design a coal cogeneration alternative. Even Du Pont, with its strong corporate policy to use coal, did not begin to investigate switching to coal at its Pontchartrain plant until existing natural gas contracts were due to expire.

Three of the firms surveyed had to make decisions as a result of air pollution requirements. An order by Pennsylvania's attorney general forced U.S. Steel to decide whether to discontinue coal use in boilers at its Homestead facility or to install electrostatic precipitators and expand coal use; the company chose the latter. General Foods discontinued coal use at its Grand Rapids, Michigan facility because the cost of air pollution control equipment necessary to meet state standards was prohibitive. Had financing been available, Johnson Paper,* a small, paper manufacturing firm, would have converted to coal because the premium for low sulfur residual fuel oil, which was required to meet air quality requirements, undermined the firm's cost competitiveness.

Industry has conducted surprisingly little systematic review of coal potential. With some notable exceptions, the industrial firms reviewed in this study did not look at their fuel-use options until they were faced with curtailments, environmental requirements, expiring contracts, or, in one case, a threat to corporate survival. Some of this behavior is understandable when we consider the confusing market signals sent to industry by public policy and the conflicting opinions of experts on the future prices and availability of various fuels. Moreover, in large segments of American industry, energy costs have been a minor competitive factor. Whatever the reasons, many opportunities for substantive cost savings were never even reviewed.

DETERMINING FEASIBILITY

The feasibility of switching to coal varies substantially with the type of facility and process. At one extreme, coal could be attractive if it was used in large boilers at spacious facilities adjacent to rail or barge transportation. At the other extreme, coal is not attractive or even feasible for use in process heaters. Beyond these two extremes, site

---

* At the firm's request, Johnson Paper is a pseudonym.

specific factors can either preclude conversions or make them economically infeasible.

Sometimes, innovation can remove technical barriers. General Shale, for example, was so anxious to move away from natural gas that it developed a new coal process. Although the brick industry originally used coal, the development of tunnel kilns, designed for cheap and then plentiful natural gas, turned the industry away from coal. By the early 1970s, with gas supply interruptions threatening, General Shale directed the efforts of its research and development staff toward adapting tunnel kilns to coal. By 1974, it had patented a "solid fuel metering and delivery system." Not satisfied with shifting all of its own plants to coal, General Shale is now urging its competitors to follow suit since it believes the industry as a whole must be strong enough to fend off competition from other building material traders.

Specific site limitations often prevent coal use. Because coal use requires a large boiler and coal handling storage and disposal facilities, the space requirements are much greater than those of oil or natural gas. Like almost any limitation, inadequate space can be overcome with enough money. But additional costs erode the financial attractiveness of coal conversion projects.

Faced with site and process limitations, firms often examine the feasibility of new or alternative technologies. Synthetic fuels could become attractive for process users that cannot use coal directly. Dow Chemical, which already has a small to medium Btu facility in the South, has been considering wider application of that technology. Exxon considered building a medium Btu facility in east Texas to supply feedstock gas supplies to industry, but has deferred plans to do so. Owens-Illinois, one of the nation's largest glass manufacturers, has explored the use of SRC-1, a solid clean coal substitute. When this study was initiated in 1980, there was substantial interest in synthetic fuels; falling real world oil prices today have considerably diminished that interest.

On the other hand, industry has expressed great interest in fluidized bed combustion. In this process, crushed coal is burned over a bed of inert ash mixed with limestone and dolomite. Fluidized bed combustors have environmental advantages, reducing sulfur and nitrogen oxides and producing a less troublesome solid waste. It is a particularly attractive technology for small boilers, where environmental controls can add greatly to unit costs. But the technology is as yet unproven for industrial use.

A few simple factors determine whether coal is feasible in any particular application. Coal's feasibility in steam boilers depends upon economies of scale, capacity utilization, and favorable site conditions. It is less feasible in process uses where flame stability and freedom from contamination are important. While these studies initially cover a range of options and approaches, they are designed to narrow the options for management consideration. Coal is the best candidate for use in boilers and kilns; otherwise, oil and gas are preferable.

PICKING CONTESTANTS: HURDLING OVER THE HURDLE RATE

Bottom-line estimates of rates of return are nothing more than a synthesis of many assumptions about the future.[4] The economic feasibility of a coal conversion project depends upon the spread between coal and oil and gas prices. This spread is extremely sensitive to the price at which natural gas equilibrates with oil prices, the level of world oil prices, and the competitiveness of the coal industry. These highly uncertain factors all represent the assumptions on which this price spread is calculated. In order to be precise in predicting the price gap, the analyst, and ultimately the decisionmaker, must implicitly predict the political stability of the Middle East, the production plans of major OPEC countries, the cost of decontrolled natural gas, future railroad rates, the potential for coal strikes, and the amount of oil and natural gas that can be extracted at reasonable prices. Since precise judgments are impossible, industrial fuel-use decisions are necessarily subject to intuitive and probably conservative judgments.

Despite what the neutral veneer analysts bestow upon their craft, it is assumptions that drive analysis--and it is people with institutional points of view that make the assumptions. Many of the firms surveyed used assumptions that supported their corporate predilections. For example, Du Pont foresaw a larger spread between coal and oil prices, creating its management's favorable view toward coal conversion. Kaiser Aluminum went even further, assuming that natural gas prices would ultimately track distillate heating oil. U.S. Steel's assumption that oil prices would rise only as fast as inflation was consistent with a general corporate policy not to project future real energy price increases.

The assumptions used will determine whether a project exceeds the hurdle rate and whether it is eligible for consideration in the capital budgeting process. Clearly, many more coal conversions would be feasible if world oil prices were predicted to increase at a constant real rate than if they were predicted to plateau or even decline. Since the fuel price differential is <u>the chief basis</u> for the anticipated rate of return, assumptions about coal, oil, and natural gas prices shape the conclusions. Expert opinion about future world oil prices has oscillated from optimism in 1978 to mild pessimism in 1979, from abject pessimism in 1980 to optimism by the end of 1981. During 1980, when pessimism was at its height, many analysts assumed that world oil prices would increase by 2 to 5 percent annually; today most forecasters predict declining real prices for a number of years, followed by stable prices throughout most of this decade. These differences in analytical assumptions about future world oil prices can dramatically change the apparent economic attractiveness of a project.

Firms can often bias calculations on different types of projects. If a firm is oriented toward developing new products, then the rate of return from such projects is likely to exceed the rate from other types of projects. A project study in one company showed that the actual present value of "cost reduction" projects was 10 percent above the level forecasted by the firm, whereas for "sales expansion" projects the actual present value was 40 percent below the forecast and for "new products" it was 90 percent below.[5] This example illustrates how firm bias can affect the outcome of analysis in favor of "productive investments." One major international energy company indicated that a "strategic" project required only a 5 percent real rate of return, while a cost-saving retrofit project would require a 15 percent real rate of return, or a four to five year payback period.[6] Since coal conversion projects are primarily for the purpose of reducing costs, rather than expanding or tapping new markets, these biases can affect their viability.

Achieving a rate of return in excess of the hurdle rate is not the only factor determining whether a project will be recommended for the capital budget. The competitive position of the firm, corporate strategy, and external stimuli, which determine whether the project is reviewed in the first place, also help to determine whether it will be recommended for final review. Since management has already been stimulated to review the project, however, most projects showing favorable economic returns will be sent forth as a capital budget request.

BITING THE BULLET

The industrial capital budgeting process contains some of the same elements as budgeting in the public sector. The most obvious similarity is that the total projects requested by various divisions exceed the capital budget, often by a considerable margin. The opprobrious term "wish list" is common to both public and private sectors. Similarly, different elements of the corporation have different perspectives on which directions to pursue. Production officials are likely to support modernization and other efforts to cut costs, while marketing officials are more interested in developing new products. The chief financial officer will protect the integrity of the firm's credit rating while staff offices dealing with environmental and energy matters are likely to support projects in their own areas of responsibility. As the project moves up the chain, alliances are formed that can be forged into a strategy.[7]

By the time a typical project reaches the capital budget stage, it has already passed the firm's hurdle rate; hence, it is already "economically justified." Now it must be compared with other projects on the basis of its contribution to corporate goals. Whether it will be approved depends on its rate of return, the amount of discretionary capital available, the number of safety and environmental projects that must be funded to comply with legal mandates, the likelihood that the facility will be closed or its operations reduced some time in the future, and the firm's priorities among cost reduction, expanding production, opening up new markets, and reducing vulnerability in the production process. The purely economic model used by most analysts calculates only the direct economic returns from the project to determine whether it will be funded.

Industry considers capital constraints a serious impediment to coal conversion. Most executives interviewed said that since competition for the capital budget was so intense, capital constraints and priorities could slow down or eliminate coal conversion projects. In theory, firms could find ways to finance attractive projects by raising more equity and debt; but this capacity was limited in the firms we reviewed. Many firms were concerned about the effects on their credit rating of increasing the ratio of debt to equity. Many did not wish to increase equity because they feared diluting their stock value. And because most firms use the capital budget as both a financial tool and a management tool in evaluating projects, total capital budget limits are assessed against organizational units in order to control expenditures and activities.

Despite these constraints, any firm can choose to raise additional capital if a project has sufficient priority. For example, two of the firms studied raised their capital budget limits--in the case of Kaiser Aluminum, by 100 percent. Nevertheless, even at this higher budget level, Kaiser's energy program would have been stretched out over much of this decade. Because of capital constraints, St. Regis stopped construction and postponed its conversion at Bucksport, Maine. As of August 1982, construction on the conversion had not yet resumed. Exxon has deferred the coal cogeneration project at its Baton Rouge refinery because of a combination of high capital costs and the current depressed state of the oil industry. In other cases reviewed in this study, the coal conversion project had sufficient priority to be funded in the existing capital budget.

Firms in the business of producing bulk intermediate products have many competing priorities. These industries--steel, chemicals, paper, and aluminum --have been forced into making large environmental investments, adding to their capital backlog. They are also extremely vulnerable to changes in the business cycle, as reduced demand for housing and consumer durables drives down demand for steel, aluminum, and building materials. Hence, just when firms in these industries need to modernize and reduce energy costs, low profits can limit their capacity to do so. Even with a strong incentive to reduce energy costs, they may have so many other modernization and environmental priorities that coal conversion projects will be dropped.

Uncertainty and even skepticism about future energy costs and availability may have a greater effect on capital budget decisions than most analysts believe. Industries have been buffeted by expert opinion from almost every direction. They were told in 1977 that natural gas was becoming increasingly scarce; a year later, a natural gas glut appeared. They were told that Middle Eastern oil supplies would be tight over the next decade and prices would continue to rise; again, a glut developed after the shock of the 1979-80 price hike hit Western economies. Already in an unfamiliar field, industrial executives are wary of making multimillion dollar decisions when the uncertainties are so great.

In many cases, conversions will be suggested at plants with indeterminate futures. In all of the industries particularly vulnerable to cyclical bursts of the business cycle and foreign competition, some plants are candidates for reduced operations or even closure. A large coal conversion investment at such a plant is a calculated risk that could backfire. Kaiser Aluminum, for example, will need to make some strategic decisions before it can make energy decisions at its Chalmette aluminum operation. General Foods decided against making additional air pollution investments at its Battle Creek, Michigan plant.

In making choices among competing projects, corporate management officials can treat coal conversion projects in three possible ways. First, they can evaluate them strictly on the basis of their rate of return. Though seldom the case in practice, this decision model is the basis for most econometric analysis and prediction. Second, they can discriminate in favor of projects that strengthen markets within their main line of business, such as developing new products, at the expense of projects designed to reduce costs, such as fuel switching. In this case, there is usually a need for some extraordinary external event to overcome the firm's general bias against cost reduction investments. Third, they can discriminate in favor of coal conversion--usually because of immediate concerns over supply curtailments (e.g., Ideal Cement and General Shale).

Various analysts have speculated about the priorities followed by industrial firms in evaluation projects. Usually firms give the highest priorities to mandatory projects, such as pollution control investments mandated by federal or state agencies and projects necessary for plant safety. Next come investments to penetrate new markets or expand existing markets, whichever is most important for a particular firm. Cost savings investments come next, with the highest priority in this category going to projects for which savings will be lost if investments are not made quickly. Run-of-the-mill cost saving projects usually receive the lowest priority, although this varies for both firm and industry. These priorities certainly bias companies against coal conversions and result in most firms using the second model above, namely, discrimination in favor of projects that strengthen markets within the firm's main line of business.[8]

THE PROCESS MODEL AND THE REAL WORLD

We are now in a position to compare how the model presented in this chapter fares in the real world. If the analysis presented in Chapter 2 is realistic, less oil would be used in small industrial boilers and very little would be used in large boilers. If reducing fuel costs was the main objective of fuel-use decisions, we should have seen large shifts during the last few years. Obviously, they have not occurred.

Low fuel costs and price controls are partly responsible for the current situation. Coal was not widely competitive with oil until 1979, and price controls still give natural gas a price advantage. And the economic recession has reduced capital investments in general. But the study uncovered coal conversion projects that were economic at even the lower world oil prices. And onc would expect a much greater level of planning and design activity than is currently occurring.[9]

The narrow economic optimization model clearly fails to describe the real world. But does the process model do any better? To answer this question, we must determine which elements have been most important in driving fuel-use decisions, and how the institutional constraints interact with economic considerations in reaching final decisions.

The first conclusion is that most U.S. firms have not felt strong competitive pressures to switch to coal. Energy costs are generally a small portion of total product costs and industrial firms feel uncomfortable diverting capital and management attention to a part of their business that does not relate to expanding markets or production. Second, the high capital costs associated with using coal raise a number of uncertainties. In order to make the investment for coal conversion at a particular site, a firm must be sure that the facility will not be jettisoned in the near future. Actual capital costs tend to be higher than estimated costs and equipment often does not run as efficiently as its manufacturers claim. And energy prices have been notoriously unpredictable. Third, because of large financial requirements for modernization and capacity expansion, firms are reluctant to tie up large amounts of capital in coal conversions. Finally, many firms lack the internal staff competence to proceed with coal conversions; thus, they must rely on outside consultants or hire new staff.

Less risky solutions are available to firms that perceive cost or security problems. Conservation is usually the quickest and cheapest way to reduce total energy costs. Overall, heavy industrial consumers reduced their demand per unit of output by 15 percent between 1972 and 1979. Because most of these improvements can be made without large capital investments, it is no wonder that industry has chosen to improve energy efficiency over fuel switching.

Obviously these observations do not apply uniformly. Ideal Cement and General Shale perceived that their future was tied to coal use. Du Pont is converting its coal-capable plants back to coal. Kaiser, Exxon, and St. Regis were all planning coal conversion projects. But even in cases where coal's prospects look more promising, we must put these actions in perspective. Kaiser, Exxon, and St. Regis have yet to make final commitments. Many firms have absolutely no intention to move to coal. And in total, industrial use of coal has decreased since the 1950s.

The process model holds up well in the real world. It does a much better job than the optimization model of explaining how and why decisions were made. But energy markets will change, and the current changes in world oil prices and tax treatment of capital investments have not yet been fully assimilated. Hence, we should examine how resilient this model is likely to be in the future.

FUTURE PROSPECTS

On the surface, many forces should strengthen coal's share of the market. Natural gas prices will increase to the level of residual fuel oil, providing great conversion potential, particularly in the South and Southwest. The new rules for depreciating investments in the 1981 Economic Recovery Tax Act create much stronger incentives to make capital investments. Generally low levels of economic activity may have hidden the actual stimulus that these changes will cause. Economic recovery could conceivably stimulate high levels of industrial investment, including coal conversion projects.

Many future coal-use forecasts are optimistic. In 1981, The Energy Information Administration (EIA), for example, projected a tripling of coal use in the industrial sector over the next decade; the National Coal Association predicted a doubling over the same time frame; and Data Resources, Inc. (DRI), more conservatively, projected a 46 percent increase by 1990. Before taking these estimates too seriously, however, one must remember that past estimates have exhibited a similar level of optimism.

These point estimates cannot hope to capture the high levels of uncertainty over the future mix of fuels in the industrial sector. Many of the forces that will shape that market are unknown today, such as the price of competitive fuels, interest rates, business reaction to changes in the tax laws, technology, and future institutional arrangements. One needs to understand the effect of these forces on the industrial energy market before the future can assume any dimension of reality.

COAL ECONOMICS AND NATURAL GAS DECONTROL

The oil price explosion after the Iranian Revolution greatly improved the economics of coal conversion. But coal's competitive position with natural gas was damaged by natural gas price controls. The major gap between controlled prices and market prices is particularly significant for coal conversions, since natural gas holds 60 percent of the large boiler market, the market for which coal is most competitive. Once natural gas is priced at its market value, coal conversion will be more attractive.

Since natural gas is most directly interchangeable with oil, its market clearing price will be equivalent to that of an oil product. A few years ago, it was generally assumed that once price controls were lifted natural gas prices would rise to the equivalent of distillate fuel oil. However, as analysts focused on the large percentage of the industrial natural gas market that competes with residual fuel oil, they have come to believe that natural gas prices will equilibrate around or below the average retail price of residual fuel oil, roughly at $4.58 per million Btu in the second quarter of 1982. Average natural gas prices for industry use at the same time were only $3.95 per million Btu, while average coal prices were about $2 per million Btu.[10]

Real coal prices are expected to rise slowly--about 1 or 2 percent annually over the next decade. Because coal reserves are so immense and access to the industry is open to large and small companies alike, competition should keep prices down. Transportation costs, on the other hand, pose a substantial uncertainty. Many utilities, for example, have complained about rapidly rising coal transportation costs and the lack of effective competition.

For a variety of reasons, natural gas prices shot up during 1982. For many distributors, natural gas prices to industry are close to or at the level of residual fuel oil prices, which have been falling. Nevertheless, in 1982, for those distribution systems charging residual equivalent prices, the gap between coal and gas was over $2.50 per million Btu on the average. On that basis, coal should represent an attractive fuel choice, even in the major producing areas. Before concluding that a major switch from natural gas to coal is imminent, however, we must remember that the economic advantages of coal over gas are no greater than coal over oil. Yet oil continues to be burned in industrial boilers where gas is not readily available or is expensive. Moreover, while gas prices are much higher, there is a widespread perception that supplies will be plentiful, which will induce some users to continue using gas.

COMPETITIVE RELATIONSHIPS

Despite the hefty increase in world oil prices and rapidly rising natural gas prices, the potential for energy costs becoming a major factor in the competitiveness of most energy-intensive industries is not great. Since energy costs represent less than 10 percent of the costs of manufacturing most major commodities, a modest saving in energy costs would not substantially change competitive relationships. Moreover, the character of certain industries makes energy costs of secondary importance. In petroleum refining, for example, access to crude oil tends to be much more important than lowering energy input costs. In the food industry, brand identification and development of new products is more important. It is necessary then to look at individual industries to determine the extent to which energy costs might become an important competitive factor.

The potential for coal use in the chemical industry is great. The industry consumes more fuel in boilers than any other, and coal use is already high--about a third of the total. Most U.S. chemical manufacturers either have or are planning to convert many of their previously coal-capable boilers back to coal. Also, most firms plan to look at coal for new boilers, although they would make decisions on a case-by-case basis. There is considerably less enthusiasm for converting existing oil or natural gas boilers to coal. However, since so many of the chemical industry's boilers are fired by natural gas, there may be a greater impetus in the future to shift to coal. Particularly in the industrial organic sector of the industry, where energy costs are high, rising natural gas prices may result in energy costs becoming a more important competitive factor. But other factors, such as rising international competition, will dwarf energy costs in importance.

The paper industry has historically faced some competitive pressures to reduce use of residual fuel oil. Many firms in the industry have reduced their reliance on residual fuel oil, mainly through greater use of wood wastes and spent pulping liquors. But there are reasons to believe that competitive pressures may play only a modest role in the future. Firms that formerly found themselves at a competitive disadvantage with residual fuel oil--compared to firms with cheaper natural gas contracts--should actually feel less pressure, as natural gas prices move up to the price of residual fuel oil. Most of the new boilers in the paper industry were designed to use oil and natural gas; conversion costs would be prohibitive in cases where many years of useful boiler life remain. Most expansion is likely to occur on existing sites, which means that spare capacity from existing boilers can be used. And fugitive dust emissions from coal handling and storage is a concern because of the fear of product contamination. At least during a period of declining world oil prices and sluggish economic activity, the paper industry is not likely to give coal conversion the highest priority, although the trend of substituting wood wastes and spent liquor for oil and gas will continue.

The potential for coal use in the petroleum refining business is high, even though very little is used today. Should Exxon move ahead on its coal cogeneration project at Baton Rouge, it might spark some interest in the industry. But the petroleum industry must weigh refinery investments against investments to find more oil and gas and upgrade refineries to produce more light products. Unless oil prices are high--improving both the economics of using coal and the cash flow of firms--coal conversion projects are likely to have a low priority. Under the conditions more likely for most of this decade, the industry will not devote much of its limited resources to coal conversion in the face of other priorities.

Neither the steel nor the food industries are likely to move to coal in any big way because of competitive pressures. The only opportunity for substantial improvements in coal use in the steel industry would be through increased production of coke, which could also displace oil or gas in a number of applications. The food industry is highly fragmented, with most boilers being relatively small. The main competitive thrust for most parts of the industry is brand identification; these firms act as highly diversified oligopolists, trying to create niches in each market they dominate. Coal conversion will not be a serious competitive concern.

Other major energy-using industries represent a mixed picture. The cement industry has already moved heavily to coal and coal is also used in lime kilns. Glass manufacturers, on the other hand, have no desire to use coal because of fear of product contamination. Energy costs are a competitive factor for Kaiser Aluminum, which faces higher costs than its competitors. The other aluminum manufacturers would probably be less interested in coal conversion than in diversifying more of their investments abroad, in countries where energy costs are low. Indeed, almost every new aluminum plant since the beginning of the 1960s has been constructed abroad.

Overall, there do not appear to be strong competitive pressures to convert to coal. With its huge potential, the chemical industry might represent the best possibility should some of the firms move aggressively ahead. It is also probable that the paper industry will attempt to move away from oil and natural gas through conversion to wood wastes and spent pulping liquors. For other major energy-using industries, the best prognosis is for continued use of oil and natural gas, except for new facilities. To the extent that energy prices affect competition or profits, the first alternative will be conservation. Coal conversion will only be used when unusual opportunities present themselves.

CAPITAL CONSTRAINTS

With oil and gas prices much higher than in 1978, the last year of relatively robust economic growth, some will argue that coal conversions should pick up once economic recovery takes hold. Some projects have obviously been deferred because of low profits and cash flow constraints. Presumably higher economic growth would stimulate capital investment in general, with coal conversion projects becoming a more important overall priority. But there are reasons to believe that these shifts will not be as substantial as pure economics would dictate. First, many of the most energy-intensive industries--such as steel, petrochemicals, and aluminum--will never recover their former position in world markets. Second, many firms will have other strong competing capital needs that have been deferred. Management attention during an upturn will probably be focused on productive investments, certain mandatory expenditures, and introducing new products, not on reducing energy costs. Lower interest rates will obviously improve the economics of coal conversion projects, but they will also improve the economics for competing projects.

Coal conversion projects are victims of rapid shifts in the business cycle. During downturns, when productive investment is less attractive, firms lack the cash flow to make coal conversion investments. During economic recovery, other capital priorities are likely to crowd them out. Coal conversion projects will be most viable during periods of substantial economic recovery, after other investment priorities have been satisfied. Hence, any shift to coal may need to wait a number of years to attain high priority in capital investment decisions.

TECHNOLOGY

Only one new technology--fluidized bed combustion --offers promise for increasing coal use. Fluidized bed combustors offer an opportunity for industrial users to burn high sulfur coal without scrubbers and to reduce disposal costs. They would be particularly valuable for use in small industrial boilers and for commercial and industrial cogeneration applications. If they were widely used, the percentage of coal consumed in small boilers could increase substantially.

Greater use of coal-fired cogeneration offers promise for many industries, such as pulp and paper, chemicals, and petroleum refining. By generating electricity and steam at the same facility, efficiencies can be improved and costs reduced. Excess electricity can be sold to utilities at their avoided production costs, creating an even stronger economic incentive. Although cogeneration might improve the economics of many projects, it adds even more to the capital costs and increases management hassle. Hence, unless third party institutions are willing to construct and operate cogeneration plants, some of the barriers facing coal conservation projects are exacerbated.

Other technologies are not likely to promote greater coal use. As mentioned earlier, coal will probably make few inroads in fueling process heaters because of its flame instability and dirtiness. Synthetic fuels, once considered so promising, are now unattractive because of much lower estimates of future world oil prices. Technical improvements in electricity technologies may actually displace some coal uses, although in general they will tend to compete mainly with natural gas. With the possible exception of fluidized bed combustors, new technology will not have substantial positive impacts on coal use.

UNCERTAINTY

Many uncertainties currently plague decisions to move to coal. The biggest single uncertainty relates to the world oil price, which has been falling in real terms since 1981. Considering the depressed demand for oil, particularly for OPEC oil, a large drop in oil prices is a real possibility and continued reductions in the real world oil price over the next few years is almost a certainty. The oil price will also affect natural gas prices, which will equilibrate at the level of residual fuel oil prices. Hence, any industrial decision to convert from oil or gas to coal must consider the possibility of lower prices undercutting the economics of the project.

At some point in the future, oil prices are likely to move upward again, improving the economic feasibility for coal conversion projects. This does not necessarily argue for making investments in coal conversion now, however; rather it argues for deferring projects until conditions are more certain.

Finally, coal transportation costs introduce another uncertainty. Even before passage of the 1980 Staggers railroad deregulation legislation, transportation costs were rising rapidly. DRI estimates that coal transportation costs in the near term will rise faster than any other component of coal utilization, eroding some of its economic advantages.[11] Recognizing that they are in a strong bargaining position, railroads have generally been unwilling to enter into long-term contracts. Hence, industry decisionmakers are caught between rising transportation costs for coal and declining costs for oil, with a substantial amount of uncertainty surrounding natural gas prices. In such an uncertain environment, there are strong pressures for delaying projects.

CONCLUSION

The current institutional structure is biased against coal conversion. Conservation and continued use of oil and gas are more attractive simply because expenditures for them generally do not have to traverse the capital budgeting process. Most firms do not systematically review facilities for coal conversion potential. Most have little competence or interest in getting into the energy business by converting to coal. The major reason for this lack of interest in coal conversion possibilities is the absence of competitive pressure to convert. In a few cases, firms have developed corporate strategies favoring greater use of coal, but other firms in the industry have not followed suit. Except for the cement industry, no one firm brought along others by moving to coal conversion as a corporate strategy.

In the competition for capital funds, investments in new capacity or other "productive" investments will take precedence over coal conversion. Firms generally expect high rates of return for coal conversion projects, reflecting some concern about the uncertainty of energy prices and the general bias of U.S. industry against longer-term, cost-cutting investments.

The prognosis for the near term is not much different. Although higher natural gas prices and better tax treatment for conversion investments should help, there still does not exist any real competitive pressure to change fuels and all of the same barriers to conversion still exist. Today, lack of cash flow inhibits coal conversion investments and has stalled the projects of Exxon, Kaiser, and Du Pont. But when economic growth picks up--creating the cash flow to make new investments--industry will feel the need to modernize, expand capacity, and introduce new products. Only a sustained recovery would provide the capital available for conversion investments.

NOTES

1. A whole body of literature, embracing the fields of psychology, sociology, management science, and statistics, exists to refute the implicit assumptions of the optimization model of industrial decision-making: that all problems are definable in terms of cost/benefit ratios, that all the information to calculate the costs and benefits is readily available in a useable form, and that decisions entail a discrete choice from among comparable alternatives. See Herbert A. Simon, "Rational Decision-Making in Business Organizations," The American Economic Review, Vol. 69,

No. 4, September 1979, for a good summary analysis of this critique.

2. That industrial firms base all capital budgeting decisions on an analysis of their competitive stance is a universal phenomenon. Michel Gouy's empirical analysis of "Strategic Decision-Making in Large European Firms," Long Range Planning, June 1978, clearly found "profitability, market share and expansion" among the top five objectives of such decisions. Strategy and Organization: Text and Cases in General Management by Hugo Uyterhoeven et al. (Homewood, Illinois: Richard D. Irwin, Inc., 1977) recommends that an analysis of the firm's position in the industry precede and serve as a basis for any major (i.e., capital allocation) decision. Michael Porter's Competitive Strategy (New York: The Free Press, 1980) also underscores the central role of strategic analysis in corporate decisionmaking.

3. Although no comprehensive study could be found on the role of energy committees or departments in corporate decisionmaking, an Archie B. Carroll and Asterios G. Kefalas study, "The Impact of Environmental Protection on Organization and Decision-Making," Managerial Planning, Vol. 27, No. 5, March-April 1979, provides an interesting surrogate. Organizational adaptation to environmental markets has produced an increase in upper level staff devoted to handling these issues. Furthermore, environmental planning has now become a part of most major corporate decisions. It is difficult to say whether energy problems hold a similar enduring quality.

4. Handling uncertainty in quantitative decisions (i.e., optimization methods) is indeed problematic. The "decision analysis" method requires very elaborate calculations of probability and assignments of risk preferences to select the best option. See Michael Menke, "Strategic Planning in an Age of Uncertainty," Long Range Planning, Vol. 12, No. 4, August 1979, and David B. Hertz, "Risk Analysis in Capital Investment," Harvard Business Review, Vol. 57, No. 5, September-October, 1979. However, empirical analysis (Lawrence D. Schall and Gary L. Sundem, "Capital Budgeting Methods and Risk: A Further Analysis," Financial Management, Vol. 9, No. 1, Spring 1980) and in-depth case studies [R. M. Cyert et al., "The Role of Expectations in Business Decision Making," in Lawrence A. Welsch and Richard M. Cyert, eds., Management Decision Making (Baltimore: Penguin Books, 1970)] suggest that numerical handling of uncertain variables is not so widespread as the prescriptive literature would suggest.

5. Joseph L. Bower, Managing the Resource Allocation Process: A Study of Corporate Planning and

Investment (Boston: Division of Research, Graduate School of Business Administration, Harvard University, 1970), cited in Richard Brealey and Stewart Myers, Principles of Corporate Finance (New York: McGraw-Hill Book Company, 1981), p. 238.

6. Teknekron, Inc. "Conversions to Coal: The Industrial Perspective," prepared for Argonne National Laboratory, Energy and Environmental Systems Division under Argonne Contract 31-109-38-6009, September 1981, sponsored by U.S. Department of Energy, Contract W-31-109-Eng-38, p. 7.

7. A number of researchers have touched on the importance of the personal aspirations and institutional role of decisionmakers in influencing decision outcomes (see books by Bower, Cyert, footnotes 4 and 5). Several writers go further, suggesting that corporate decisions may actually represent negotiated compromises among various decisionmakers involved; unfortunately, little information is offered as to the basis of these compromises. See Andre Delbecq, "The Management of Decisionmaking Within the Firm: Three Strategies for These Types of Decisionmaking," The Academy of Management Journal, Dec. 1967, Vol. 10, No. 4, pp. 329-339, and Ross Stagner, "Corporate Decisionmaking: An Empirical Study," Journal of Applied Psychology, Vol. 53, No. 1, February, 1969.

8. Teknekron, Inc., Argonne National Laboratory, p. 7.

9. Coal boilers have captured a large share of new sales, reflecting the improved economics since 1979. But one would expect a much faster rate or expansion of total sales if any substantial number of existing facilities were switching to coal.

10. Alvin L. Alm and Frank Clayton Schuller, "Regulating Rigidity: Turmoil in Natural Gas Markets," (Working Paper, Energy and Environmental Policy Center, John F. Kennedy School of Government, Harvard University, February 1983).

11. Data Resources, Inc., Coal Review Update, Summer 1982.

4
The Clean Air Act and Industry Decisions

The Clean Air Act of 1970 was passed just as the extent of the U.S. energy crisis was becoming obvious. The peaking of crude oil production in 1970, coupled with rising demand, set the stage for an 83 percent increase in oil imports over the next three years. But in 1970--the year of Earth Day, the creation of the Council on Environmental Quality (CEQ) and the Environmental Protection Agency (EPA), and President Nixon's comprehensive environmental program--environmental quality ranked as one of the nation's major domestic priorities, while concern about energy was in a nascent stage. The symbol of smog-ridden Los Angeles pushed air pollution control to the forefront of the 1970 legislative agenda.

The Clean Air Act was the first major environmental legislation enacted in the new decade. Similar to the previous Air Quality Standards Act of 1967, it depended upon ambient air quality standards to protect public health. This time, however, Congress set deadlines for achieving the standards. Establishment of emissions standards for new sources of pollution was another innovation that became the mainstay of future air and water pollution legislation. Although the automobile was the major public focus of the 1970 legislation, the new source standards and legislative deadlines created a new and much more effective framework for stationary pollution control efforts.

If the 1970 Clean Air Act established the skeleton for the nation's air pollution program, the 1977 amendments created the teeth. The legislation dealt with two simple issues. First, recognizing that the 1970 Act's deadlines had not been achieved, Congress created new deadlines and a procedure to implement them. Second, recognizing the Supreme Court's injunction for EPA to protect clean air areas, Congress essentially codified into law EPA's regulatory response to the Court's mandate. By prescribing specific emission reduction levels and deadlines for compliance, the 1977 amendments tightened the congressional hold on the air pollution control program.

Existing standards were to be achieved by a wide array of new regulatory controls, designed to curb emission growth in areas not achieving the health-related air quality standards. No new growth was to be permitted after July 1, 1979, unless a state provided for achievement of the ambient standards by December 1982, or, under certain circumstances, 1987. During the interim, states had to assure that new growth was accommodated by more than offsetting emissions from other sources, either within the same facility or from some other facility. The new source also had to achieve the Lowest Achievable Emission Rate, that is, the best level of emission control ever achieved for the particular polluting process. Theoretically, had 98 percent removal of a certain pollutant been achieved at some place in the country by using a particular technology, that technology would also have to be applied to a new facility in a non-attainment area.

The new controls aimed at clean air areas were perhaps even more complex. The Prevention of Significant Deterioration (PSD) provisions were aimed at protecting already clean areas through the establishment of a three-tiered zoning-permit system, similar to the system established by EPA regulations in December 1974. The 1977 amendments designated all national parks, wilderness areas and other special natural attractions as Class I areas. All other areas of the country were designated as Class II, allowing states to redesignate Class II areas as Class III areas. None has yet done so. The PSD program held emissions in Class I areas to roughly 2.5 percent of the annual ambient air quality standard, Class II areas to roughly 25 percent, and Class III areas to 50 percent.

In addition to the zoning system, the Clean Air Act required use of Best Available Control Technology (BACT) in all clean air areas. BACT was designed to be more stringent than the New Source Performance Standards (NSPS) already required by the 1970 amendments. In practice, however, BACT has generally been identical to the NSPS. (For a more detailed discussion of the Clean Air Act requirements, see Appendix A.)

The requirements in both dirty and clean areas are heavily aimed at coal combustion. Use of coal produces sulfur oxides, particulates, carbon monoxide, and nitrogen oxides, as well as trace amounts of radionuclides and heavy metals. Because of the heavy polluting nature of coal, it was inevitable that, next to the automobile industry, coal combustion would be most directly affected by the Clean Air Act.

Conflicts between energy and environmental goals first emerged when Congress considered legislation that ultimately became the Energy Supply and Environmental Coordination Act of 1974 (ESECA). The most prominent feature of that legislation was federal authority to mandate coal conversion of power plants. After painful deliberation between the Senate Interior and Public Works Committees (now the Energy and Natural Resources and Environment and Public Works Committees), a compromise was struck. The Federal Energy Administration was given authority to mandate conversions as long as ambient air quality standards could be achieved, thus suspending the New Source Performance Standards that normally would have applied to major modifications. EPA was given responsibility for determining whether emissions from potential coal conversion candidates would violate the ambient air quality standards.

During debate on the 1977 amendments, industry objected that the PSD provisions would prevent siting of industrial and utility plants throughout large parts of the country, and that the costs of BACT would be prohibitive. The Carter Administration asserted in its National Energy Plan that a conflict did not exist between energy and environmental objectives. After considering the Clean Air Act for almost three years, Congress was ready to act and overwhelmingly passed the 1977 amendments.

THE SECOND ROUND OF AMENDMENTS: THE ISSUES EMERGE

After passage of the 1977 amendments, many industry groups were extremely concerned about the negative effects the Clean Air Act would have on industrial development and fuel-use choices. Industry groups charged, for example, that the Act would be prohibitively costly, would result in delays, would divert capital into unproductive investments, and would delay and kill projects. This chapter deals with these assertions as they would affect the use of coal. It reviews how higher costs imposed by environmental requirements affect coal conversion economics, the impact of capital diversion in drying up capital sources, and the impact of delays and uncertainty on industrial decisions to move to coal. It then probes the strategies industry would employ to overcome potential limitations arising from the Clean Air Act. It concludes that the Clean Air Act has not been a major barrier to industrial coal conversion. Even in the future, when the PSD limitations become more binding as sources are added, the Clean Air Act would most likely inhibit coal conversion in only a few regions.

THE COSTS OF POLLUTION CONTROL

Air pollution control is expensive. Electrostatic precipitators, required on most facilities, add approximately 10 percent to the capital costs of a facility and a much smaller portion of operating costs. If scrubbers were required for industrial boilers, they could add as much as 33 percent to capital costs, over 100 percent to operating costs, and perhaps 25 percent to the overall cost of using coal.[1] One must be very cautious in using average costs such as these, however, since actual environmental costs will depend heavily on the type of equipment used, a variety of site specific factors, boiler size and capacity utilization.

Use of stack gas scrubbers has never been a generic requirement. States usually apply emission limits to all boilers, both new and existing, but allow use of low sulfur coal to achieve them. Federal NSPS have never applied to small industrial boilers and EPA has not required use of stack gas scrubbing for industrial boilers. Should the NSPS for industrial boilers require scrubbers for new coal-fired boilers in the future, however, EPA estimated that the demand for coal use in new boilers from 10-50 mmBtu/hr would fall from 47 to 6 percent, and in boilers from 50-100 mmBtu/hr from 58 percent to 16 percent. For larger boilers, however, from 100-250 mmBtu/hr, EPA projects the drop would be from 74 to 65 percent and above 250 mmBtu/hr, coal penetration is virtually unaffected.[2]

Chapter 2 indicates similar but less negative results. With the mid-price oil levels, coal use in small boilers (70 mmBtu/hr) would be marginal, except where facilities are near low cost Rocky Mountain coal supplies and are operated at high capacity. As the size of facilities increases, the economics improve, although gradually. In both large and small boilers, high capacity is the primary factor that makes coal an economic investment; not a surprising result, considering that scrubbers add so substantially to unit costs at low capacity levels. The much more robust economies of scale and more stringent impacts found in the EPA analysis--compared to this analysis--are partially explainable by the fact that this study reflects the generous depreciation benefits of the Economic Recovery Tax Act of 1981, which has the effect of substantially lowering capital costs. The extent of these differences, however, suggests further analytical differences.

The Clean Air Act and state requirements do stimulate greater demand for low sulfur coal which is more costly. The differential between low and high sulfur coal was about 30 percent at the minemouth for Appalachian coal in 1982. The difference between low and medium sulfur coal, on the other hand, was about 10 percent. By the time transportation, capital, and operating costs are taken into account, the extra cost for burning low sulfur coal represents at most an addition of 3 to 6 percent to delivered costs.

Permitting represents another significant cost. In a study of air quality permitting, Environmental Research Technology (ERT), Inc., estimated that the average direct cost of permit applications up to submission was $250,000 to $300,000.[3] For a small facility, permitting costs can represent a substantial share of project costs and a formidable paperwork barrier. For a large project, permitting costs are generally less imposing as a percentage of total project costs.

The case studies and interviews we conducted revealed few situations in which the costs of meeting air quality requirements had a significant impact on the industrial fuel-use decision. Only one company abandoned use of coal because pollution control expenditures would have been too costly. General Foods switched from coal to natural gas in three small boilers with multi-fuel capability at its Battle Creek, Michigan facility, because the cost of particulate controls was considered prohibitive. Installation of a baghouse required by the state was not considered costeffective, given the relatively small fuel savings associated with the use of coal in its small boilers. American Cyanamid was dissuaded from switching to coal when a preliminary study at one of its plants indicated that imposition of a stringent BACT requirement was possible. But no detailed economic analysis was conducted. In all other cases, coal was economic even with the application of particulate controls and the use of low sulfur coal, buttressing the conclusions in Chapter 2 that, at least with high oil prices, coal has substantial economic advantages over oil.

CUMULATIVE CAPITAL COSTS: THE BARE CUPBOARD

Total pollution control expenditures, by constraining capital budgets, could also indirectly preclude coal conversions. Table 4.1 shows that total pollution control capital expenditures from 1978-88 would be $375 billion; equivalent to about 1 percent of the Gross National Product (GNP) estimated for that period. For 1979 alone, air pollution industrial capital costs were almost $12 billion, roughly one half of the total capital costs for all environmental requirements.

A more careful review, however, softens the initial impression. The Commerce Department's Bureau of Economic Analysis estimates that pollution control investment equaled only 3.1 percent of total investment in new plants and equipment in 1979 and 1980, and only 2.8 percent of total plants and equipment in 1981--surprisingly low figures[4] (see Table 4.2). For most major energy-using industries, environmental capital investments are less than 10 percent of total capital costs. In many industries with high environmental

TABLE 4.1

ESTIMATED TOTAL POLLUTION CONTROL EXPENDITURES, 1979-88
(billions of 1979 dollars)

	1979			1988			Cumulative 1979-88		
	Operation and Maintenance	Annual Capital Costs[1]	Total Annual Costs[2]	Operation and Maintenance	Annual Capital Costs[1]	Total Annual Costs[2]	Operation and Maintenance	Capital Costs[1]	Costs[2]
Pollutant/source									
Air Pollution									
Public	1.7	.4	2.1	2.8	.7	3.5	22.5	5.3	27.8
Private									
Mobile[3]	3.2	4.9	8.1	3.7	11.0	14.7	32.1	83.7	115.8
Industrial	2.5	2.9	5.0	3.9	5.1	9.0	32.5	41.5	74.0
Utilities	6.3	3.5	9.8	8.5	6.5	15.0	71.1	50.1	121.2
SUBTOTAL	13.7	11.7	25.4	18.9	23.3	42.2	158.2	180.6	338.8
Water Pollution									
Public	3.7	8.2	11.9	5.5	14.2	19.6	45.4	99.7	145.1
Private									
Industrial	4.4	3.2	7.6	6.4	5.1	11.5	52.4	41.1	93.5
Utilities	.4	.5	.9	.4	1.1	1.5	3.6	7.8	11.4
SUBTOTAL	8.5	11.9	20.4	12.2	20.4	32.6	101.4	148.6	250.0
Solid Waste									
Public	1.7	.3	2.0	2.5	.6	3.1	21.8	5.3	27.1
Private	4.5	.7	5.2	7.5	.6	9.1	61.3	12.5	73.8
SUBTOTAL	6.2	1.0	7.2	10.0	2.2	12.2	83.1	17.8	100.9
Toxic Substances	.1	.2	.3	.5	.6	1.1	3.6	4.6	8.2
Drinking Water	.3	.4	.7	.7	.8	1.3	5.3	5.3	10.5
Noise[3]	<.05	.1	.1	.6	1.1	1.6	2.6	4.3	6.9
Pesticides	.1	<.05	.1	.1	<.05	.1	1.6	.1	1.7
Land Reclamation	.4	1.3	1.7	.4	1.4	1.8	4.5	13.5	18.0
TOTAL	29.3	26.6	55.9	43.2	49.7	92.9	360.3	374.7	735.0

[1]Interest and depreciation.

[2]Operating and maintenance costs plus capital costs.

[3]Incremental and total costs are assumed to be the same.

Source: Council on Environmental Quality, Environmental Quality--1980, December 1980, p. 397.

TABLE 4.2

INVESTMENT FOR POLLUTION CONTROL BY U.S. INDUSTRIES
(billions of dollars)

	1981 Pollution Abatement				
	Air	Water	Solid Waste	TOTAL	Percent of Investment[2]
All Industries[1]	4.97	3.04	0.92	8.93	2.77
MANUFACTURING	2.69	2.10	.63	5.42	4.27
Durable goods	1.09	.70	.18	1.97	3.18
Primary metals	.54	.19	.05	.78	9.60
Blast furnaces, Steel works	.33	.13	.02	.49	15.46
Nonferrous metals	.16	.05	.03	.23	6.65
Fabricated metals	.02	.04	A	.07	2.36
Electrical machinery	.08	.07	.02	.18	1.75
Machinery, except elec.	.05	.09	.01	.15	1.13
Transportation equip.	.20	.21	.06	.46	2.50
Motor vehicles	.16	.16	.04	.35	3.47
Aircraft	.03	.05	.02	.10	1.56
Stone, Clay, & Glass	.12	.03	.01	.16	5.10
Other durables[3]	.07	.07	.02	.16	2.81
Nondurable goods	1.60	1.40	.45	3.46	5.33
Food including beverage	.13	.14	.04	.30	3.65
Textiles	.03	.02	A	.05	3.20
Paper	.16	.12	.11	.38	5.65
Chemicals	.38	.36	.14	.88	6.47
Petroleum	.88	.74	.14	1.76	6.89
Rubber	.02	.02	.01	.04	2.26
Other nondurables[4]	.02	.01	.01	.04	.61
NONMANUFACTURING	2.28	.94	.29	3.51	1.80
Mining	.18	.18	.10	.46	2.72
Transportation	.04	.04	.01	.09	.75
Railroad	.02	.02	A	.04	.94
Air transportation	.01	A	A	.01	.26
Other transportation	.02	.02	A	.05	1.25
Public Utilities	1.98	.67	.15	2.80	7.29
Electric	1.91	.65	.15	2.71	9.11
Gas and other	.06	.03	A	.09	1.04
Trade and Service	.05	.04	.03	.11	.13
Communication, commercial, and other[5]	.02	.01	A	.03	.07

[1]Excludes agricultural business; real estate operators; medical, legal, education, and cultural services; and nonprofit organizations.

[2]Pollution control as a percent of total plant and equipment investment.

[3]Consists of lumber, furniture, instruments, and miscellaneous.

[4]Consists of apparel, tobacco, leather, and printing-publishing.

[5]Consists of communication; construction; social services and membership organizations; forestry, fisheries, and agricultural services.

A: Less than $5 million.

Source: Bureau of Economic Analysis, U.S. Commerce Department Survey of Current Business, June 1982.

investments, such as petroleum refining or chemicals, cash flow has been acceptable over the years and investment requirements have not been oppressive. The steel and non-ferrous metals industries, however, have high pollution control investment requirements, while facing low demand and profits. The capital costs associated with pollution control have reduced discretionary expenditures in these industries for new capital investment, which could affect the capacity of the steel and aluminum industries to convert to coal. But considering the multiplicity of competitive and other problems plaguing these industries, it would be almost impossible to determine what capital decisions would have been made if pollution control investments had been smaller.

COSTS FROM OTHER LAWS

Air pollution requirements are not the only ones facing the industry. The Clean Water Act is similar in both ambient and technological standards, and deadlines for compliance. The estimated industrial capital cost requirements are somewhat lower than those for the Clean Air Act. These costs are somewhat understated, however, since much industrial waste is treated in municipal plants. Other major regulations such as those contained in the National Environmental Policy Act, the Fish and Wildlife Coordination Act, and the River and Harbor Act, affect the siting and development of industrial projects. Despite the real impact of all these environmental laws on industrial actions, none were mentioned by industry respondents as significant factors affecting fuel-use decisions.

Regulation of solid waste under the Resource Conservation and Recovery Act (RCRA), on the other hand, was mentioned as a source of some uncertainty. Initially, EPA proposed to treat fly ash, scrubber sludge, and other products of fossil fuel combustion as "special waste," a category that recognized the need to treat bulky, slightly hazardous wastes differently than highly toxic chemicals.[5] The 1980 RCRA amendments suspended regulation of those wastes entirely.[6] They required EPA to conduct a study of the effects of these wastes within two years.[7] As of the date of this publication, EPA has not completed a review of fly ash, scrubber sludge, and other products of fossil fuel combustion. Apart from the uncertainty inherent in the current suspension of regulation of these wastes, there is also considerable uncertainty about the ultimate form and stringency of regulation under RCRA.

EFFECTS OF DELAYS

The time required to obtain environmental permits affects firms differently. The lead time for constructing and installing a large utility boiler from initial concept to startup is generally six to eight years, even without any additional time for environmental permitting. The lead time for smaller package boilers and fuel-use projects, on the other hand, may be reasonably short--one to two years. When environmental permitting is integrated into the planning process, firms may experience no actual delays. On the other hand, when a firm lacks experience with the complexity of permitting and does not build in sufficient lead time during the planning process, substantial delays may result. Even experienced and sophisticated firms may encounter unusual problems or bureaucratic delays in obtaining some permits, which lengthen the permitting process beyond normal planning timetables and cause construction delays. Firms desiring to convert small boilers may experience delays in bringing them on-line, simply because the length of time required to obtain environmental permits approaches that of installing a pre-packed boiler.

Many companies reported minimal problems with project delays due to air quality permitting problems. General Motors, for example, had applied for approximately twenty PSD permits over the course of three years and had experienced no great difficulties in obtaining permits. Most of GM's applications had been acted upon within three to four months, with an occasional permit requiring six months. On the other hand, a very small paper firm received timely review of its permit only by daily attention to the permit's progress through the review process. Even where companies have experienced some problems with delays due to air quality permits, as General Shale did in its Chattanooga plant, the overall sense is that coal conversions and expansions using coal need not be subject to significant delays due to permitting problems.

Some firms feared time delays even though they had not experienced any. American Cyanamid had considered applying for a Department of Energy (DOE) prohibition order under the Fuel Use Act to facilitate conversion of plants in New York State, but was dissuaded from doing so because of the unfavorable experience a local utility was having in gaining its permits.

The gap between quantitative data on delays and industry perception is large. The National Commission on Air Quality concluded that 75 percent of PSD permits were issued between three-and-one-half and ten months, and only 25 percent took between ten and sixteen months.[8] These data however did not include time estimates for monitoring requirements, which can take up to a year, nor requests for additional information before a permit can be officially received. A number of the firms interviewed felt that permits could and would take two to three years to be issued, when all of these steps were included.

Project delays caused by the permitting process have two major effects. The first and most obvious is to increase the cost of the project if the permit is on the critical path. For many firms, the second effect--the loss of their competitive edge--is a more serious problem. If a firm is attempting to get a new product on the market, the last thing it wants is a long delay that will allow its competitors time to develop a counter-strategy. In highly competitive markets, delays can easily turn a profitable project into an unprofitable one. Proctor and Gamble has experienced permit delays at its Port Ivory facility in Staten Island, New York, and at its Long Beach facility in California. Although the delays did not lead to project cancellations in these particular cases, P&G remains very concerned about the impact of permitting delays on projects in highly competitive markets. Other firms expressed similar concerns about the importance of getting products to market quickly, before their competitors could take advantage of their plans.

While more experience with the PSD process might, over time, speed implementation, greater demand for permits created by increased economic growth and greater pressure on increments may create new impediments. As long as a permit-by-permit PSD provision is retained, some delays unsatisfactory to applicants are almost unavoidable, leading to site-hopping and reconfiguration of some projects. For this reason, as well as other more compelling reasons, the current trend toward industrial expansion at existing sites will likely continue.

UNCERTAINTY

One premise underlying this study was that uncertainty might play a large role in inhibiting coal use. Concern is often expressed that should a coal conversion project be undertaken, the opportunities for regulatory ambush or rule changes represent almost permanent peril. Faced with at least the possibility of a catastrophic outcome, firms might choose to continue using oil or natural gas, which are more acceptable to the regulatory agencies.

The evidence does not support the contention that uncertainty is a major impediment to coal use. For the most part, the environmental staffs of companies understand the environmental requirements their firms will have to meet and believe that they will not face any undue problems in securing permits. Most of these fairly confident attitudes arose in the Sunbelt states of the south and southwest, where state environmental officials are considered more cooperative and the state governments clearly encourage industrial development.

The environmental agencies in states such as Texas and Louisiana, for example, have long enjoyed a reputation for friendly and cooperative attitudes toward business. These attitudes are significant because so much of the country's manufacturing growth has occurred in these areas, particularly in petroleum, petrochemicals, and chemicals. In states such as California and Washington, however, environmental officials are considered less accommodating and some firms encountered problems in attaining prompt environmental approvals. In almost all cases, industry environmental staffs claimed that they had run into problems in dealing with EPA, particularly at the headquarters level. However, EPA generally was considered less a barrier in accommodating states than state agencies along the West Coast.

Industry was not concerned about retroactive environmental requirements. Environmental staffs in the large manufacturing firms know that, at least on the basis of past experience, retroactive requirements are unlikely. Industry officials were also generally confident that states would forgive equipment failure.

The type of project determines the amount of uncertainty encountered. A multistate pipeline with terminal dock facilities is likely to require dozens or maybe even hundreds of individual permits, making these projects potential examples of environmental horror stories. For example, the SOHIO pipeline in California required 700 permits, and the proposed Northern Tier pipeline from Puget Sound to Minnesota has encountered a host of regulatory difficulties. Proposed refineries in Hampton Roads, Virginia, and Eastport, Maine, have also encountered delays. An Exxon drilling project off Santa Barbara, California, was delayed by permit requirements for a number of years. A simple expansion of capacity in a state like Louisiana represents the other extreme, where industrial environmental officials can predict fairly accurately the disposition of a permit application. While the dramatic examples are likely to gain public attention, most coal conversion permits appear fairly straightforward and routine. In the southern Sunbelt areas, gaining environmental permits might be the least of a firm's problems with coal conversion.

Permitting new sites quickly is much more difficult than permitting existing ones. New greenfield sites require more modeling and monitoring than existing facilities and are likely to require additional permits for water discharges, dredging and filling. In some states, California the most notable example, developing a new site can involve a myriad of permits from various state and local agencies. Other states, Washington being the most obvious example, have condensed multiple permitting requirements into a one-step process, although the state has rigorous environmental review. Developing greenfield sites not only involves greater environmental risk and cost, but also requires much greater outlays for infrastructure development. Hence, environmental requirements represent one reason so much expansion occurs at existing sites.

For those firms that need to develop greenfield sites in order to be closer to markets, environmental requirements have led to a certain amount of site hopping. One firm, for example, developed several site options for a new facility, allowing it the luxury of choosing the site least likely to cause permit delays. Because of real or perceived difficulties in gaining permits, many firms shy away from developing new sites, particularly firms concerned about exposing their strategy to competitors.

Uncertainty is sometimes a synonym for inexperience. Firms such as Du Pont and General Motors, with substantial experience burning coal, have few problems with uncertainties in the permitting process. Smaller firms, undertaking their first coal conversion, perceive rather difficult obstacles. From our study, however, uncertainty did not deter small firms from deciding to move to solid fuels, although the firms did encounter greater problems in obtaining permits.

For most firms, uncertainty over air pollution permits pales in significance compared to other uncertainties they must face. Although the firm has no guarantee how quickly it will gain a permit, the total variance is much smaller than many business unknowns, such as the future rate of economic growth, inflation, interest rates, the potential market for their product, and the likely counter-strategies of their competition. For the firm with experience, uncertainty, if anything, should stimulate more coal use, since subsequent resolution of the uncertainty could result in more restrictive requirements and depletion of clean air increments by other emitters. Although few firms have resolved uncertainty by pushing ahead, uncertainty has probably not hindered many coal conversion decisions either.

THE CLEAN AIR ACT AND COAL USE: IS THE PAST PROLOGUE?

Based on the evidence from this study, environmental quality requirements have not yet had a major impact on fuel-use decisions. The General Foods example represented the only documented case where environmental costs were so high that the firm switched to natural gas. Fear of delays obviously changed some decisions, such as American Cyanamid's decision to forego requesting a DOE prohibition order that would have eliminated the need for a formal federal PSD review because of a utility's experience with delays. An Argonne National Laboratory report to DOE mentioned "an isolated incident in which a big plant was not built in southeastern Texas because of 'environmental' pressures."[9] A Goodrich chloralkali plant, which would have burned coal, changed its plans to build in Texas after resistance by consumer and environmental groups, despite receiving the necessary permits and agreeing to use scrubbers. The plant was subsequently built in Louisiana. But with this handful of exceptions, PSD permits were being processed without inordinate delays and internal offsets were generally available for conversions in non-attainment areas.

It is not surprising that the Clean Air Act, particularly the new PSD and offset provisions, has had such minor impact on coal use. Since the PSD provisions were designed to budget clean air increments far into the future, one would not expect many increments to have been used up yet, particularly considering that the 1977 amendments are relatively recent in origin.

But the future could be different. A study conducted for the Business Roundtable concluded that potential fuel switches from natural gas to oil or coal in Texas could constrain development on the Gulf Coast, and foresaw PSD limitations on use of coal in areas of rough terrain.[10] It is also likely that industry is currently using up the cheaper internal offsets, leading to prospective scarcity and higher prices for general offsets in the future. There have already been isolated reports of plants unable to secure necessary offsets at almost any price. The current ability of most firms to find suitable offsets may result from the relatively low level of new investment in plants and equipment and excess steam capacity and thus reflect an abnormally low demand for offsets. Once economic activity picks up, demand for offsets and PSD permits could grow. Moreover, as firms exhaust their housekeeping conservation potential, industrial capacity will not be able to expand without some growth in steam capacity.

POTENTIAL INDUSTRIAL RESPONSES

Most industrial firms interviewed for this study expressed concern about the future availability of both offsets and PSD increments. Since ultimately some of these firms will actually face permit barriers from either PSD or non-attainment requirements, we were interested in how they would react to these concerns at particular sites. A range of options are available to them--from abandoning the project to challenging the regulators. They can choose to use a cleaner fuel or fuel mixture or perhaps purchase steam or electricity from an external source. They can choose to reduce potential emissions to avoid EPA review. They can offset the proposed emission increase with a reduction in existing emissions or purchase offsets from another firm. Finally, they can change the size of projects or even abandon them.

In reviewing options that they might employ to cope with a Clean Air Act restriction, industry representatives displayed a number of general concerns. Firms were almost universally reluctant to sell offsets at their facilities and had little faith in their ability to purchase offsets from other firms. Most firms responded to general questions of probable future constraints posed by PSD limitations and offset unavailability by expressing concern over the impact of air quality restrictions on future growth. However, when faced with a specific hypothetical situation involving expansion at current plant sites or future sites, every respondent indicated several ways in which the firm felt it could overcome the apparent air quality restriction to complete project plans.

Faced with a hypothetical inability to gain a PSD permit, most firms indicated they would challenge the modeling results. This response reflects a common concern that the models are overly conservative, a criticism that EPA accepts as just cause for changing PSD permit determinations. Union Carbide officials, for example, asserted that the first thing they would do if the EPA model showed increment violations would be to challenge the model. Because short-term models combine worst-case emission rates and worst-case meteorological conditions, air pollutant concentrations often are predicted to be higher than actually occur. Union Carbide used actual monitoring data to dispute modeling results in a case in Texas that showed model results exceeded actual concentrations by 400 percent.

One company maintains "highly qualified" air quality modeling consultants to challenge the "very conservative" models. Should PSD increments impede future plans, the firm might use the consultants to dispute model results. Houston Power and Light, a Texas-based utility, also indicated that, should other negotiations fail, the air quality models would be challenged.

A variation on challenging the model in PSD areas is to challenge the BACT determination of an earlier PSD applicant. A later applicant can thus force adoption of a more stringent technology by a neighboring plant. This strategy can "create" a PSD increment even where a state agency had previously determined there was none.

Next to challenging models, firms indicated that they would use offsets either to make the project "de minimis" (below the size requiring an air quality permit) or to reduce net emissions to allowable levels. Since the offset market is limited by the few companies willing to sell offsets, most firms felt that the possibility of obtaining external offsets could be quickly dismissed as an option. In almost all cases, firms proposed to offset emissions from their own operations. Should they be available, however, firms would be interested in purchasing them, assuming the price was not exorbitant. Union Carbide, General Motors, and Gulf State Utilities all claimed that they would consider buying offsets if available at a reasonable cost.

For both PSD and offset constraints, firms were split on the next likely options they would pursue. Two firms voiced a preference for switching sites. Other tactics mentioned by firms included attempting to achieve redesignation of an area's attainment status and using political clout to get a favorable permit determination.

In only one case did a firm admit it would rush through a PSD permit to be the first to obtain an available increment. Changes in plant design, fuel switches, and addition of better pollution control technology were generally ranked well below the tactics discussed here. Only Union Carbide staff displayed a willingness to change plant decisions, by switching to cleaner fuels or installing additional pollution control technology. They were, however, adamant that they would not change the size of a project to accommodate air quality constraints; they would prefer to change sites.

The themes woven through the industrial response about future constraints may be more useful to future energy and air quality policy decisions than the actual industry responses. The commonality of ideas, fears, and hopes expressed by industry groups indicate areas for improved policy and communications.

First, as mentioned, a startling discrepancy existed between firms' perception of problems and their actual experience. This apparent dichotomy is visible throughout the interviews with industry. The best explanation seems to be the difference between industry officials' ability to grasp the tangible versus the intangible. When confronted with a problem similar to that already encountered during plant operations, industry officials turn to the channels that they have used in the past. On an individual plant basis, something can usually be done to resolve problems. However, when discussing reform of air quality regulations in the abstract, the problems appear insurmountable and constraints on industrial growth are perceived as serious.

Second, most firms question the viability of a publicly operated offset market, particularly because they were so reluctant to sell their own offsets. Of the thousands of recorded offset transactions, only a small portion occurred between firms; the rest occurred within the same company. If this pattern of holding offsets for potential future expansion persists, it could result in a much higher value for the offsets available and in even greater shortages.

Some firms are troubled by uncertainties over the availability of PSD increments and offsets for both clean and dirty air areas. It is virtually impossible to pinpoint where future economic growth will take place, which industries will grow the fastest, what technologies they will use, and hence to what extent, if any, PSD will act as a limit to industrial growth. It is clearer that in response to such limitations, industry will first challenge the regulators--either the modeling results or in case of non-attainment areas, the designation.

It is less clear what additional industrial strategies will be available and workable when regulatory limits are reached. If virtually all limits are challenged and industry wins most of these challenges, the limits imposed by the 1977 amendments would be less restrictive than envisaged. Assuming the law means what it says, however, the PSD limits will be binding as increments are used up and growth in non-attainment areas will be restricted without offsets. In some cases, the simplest answer will be to use the cleanest fuels available, either natural gas or low sulfur oil.

This brings us to a final question. Even if environmental laws have not constrained coal use in the past, should we, from an energy point of view, be concerned about likely future restrictions, particularly in the ability to shift facilities to coal in fast-developing areas? Are the concerns that spawned ESECA and the Fuel Use Act still relevant in today's environment? These issues will be discussed in the final chapter.

NOTES

1. Data Resources, Inc., Coal Review Update, Winter 1981-82, p. 60.

2. Environmental Protection Agency, Impact Analysis of Alternative New Source Performance Standards for Industrial Boilers: Energy, Environmental, and Cost Impacts. Preliminary Draft, October 1980, pp. 7-11 to 7-14.

3. The Business Roundtable Air Quality Project, Volume III, The Impact of Air Quality Permits Procedures on Industrial Planning and Development, prepared by Environmental Research and Technology, Inc., Cambridge, Massachusetts, November 1980, p. 52.

4. U.S. Department of Commerce, Bureau of Economic Analysis, Survey of Current Business, Volume 62, No. 6, June 1982, p. 18.

5. See 43 Fed. Reg. 58991-91 (Dec. 18, 1978).

6. Solid Waste Disposal Act Amendments of 1980 (SWDAAm), §7, Pub. L. No. 96-482 94 Stat. 2337. The exemption of these wastes from regulation, pending completion of study and further rulemaking, is contained in 40 C.F.R. § 261.4(b)(4).

7. SWDAAm §29.

8. National Commission on Air Quality, To Breathe Clean Air (Washington, D.C., 1981), pp. 21-22.

9. Teknekron, Inc. "Conversions to Coal: The Industrial Perspective," prepared for Argonne National Laboratory, Energy and Environmental Systems Division under Argonne Contract 31-109-38-6009, September 1981, sponsored by U.S. Department of Energy, Contract W-31-109 Eng.-38.

10. The Business Roundtable Air Quality Project, Volume II, The Effects of Prevention of Significant Deterioration on Industrial Development, prepared by Arthur D. Little, November 1980, p. v.

5 Energy and Environmental Quality

The historic policy interest in coal conversion--codified into law in the Fuel Use Act--arose from two concerns. First, greater coal use was considered a way to back out imported oil, which, until recently, posed a threat to national security. Second, coal conversions would also reduce dependence on natural gas, which was considered a dwindling resource. Today, oil imports are low, OPEC surplus capacity is 50 percent of production and real oil prices are falling. Natural gas wells are being shut-in and reserve additions in 1981 exceeded production. These changes raise obvious questions about the current and prospective importance of coal conversion to energy policy--questions that seemed highly improbable just a few years ago.

The earlier discussion on industry's reluctance to use coal raises a fundamental question about how energy services are provided. If industrial firms face conflicting priorities in allocating funds to economic coal conversion projects, would other entities find such activities a profitable line of business? For example, a pulp and paper company might find its capital budget too constrained to fund a coal conversion project. But an electric utility or boiler manufacturer might conclude that building and operating a coal-fired cogeneration plant represented an investment preferable to its other alternatives. Creation of new entities to finance and operate industrial boilers and cogeneration facilities might do much more for coal conversion than reducing pollution control requirements.

This concluding chapter focuses on both issues. It reviews the role coal conversion plays in energy policy today--as compared to the role it appeared to play in the past. The chapter then examines the industry structure for providing steam and electricity, concluding that a fundamental shift in structure could be important for increasing coal use.

ENERGY GOALS

Two concerns drove the public policy impetus behind coal conversion. First, after the Arab oil embargo, greater use of coal appeared to represent a major opportunity to reduce oil imports. As the most abundant U.S. domestic resource, there were often fanciful dreams of reaching or at least approaching independence from imports. Second, up until 1978 there was a pervasive feeling that natural gas supplies were becoming increasingly limited; conversion to coal in industrial and utility boilers would save this premium fuel for residential and commercial use, feedstocks, and other uses where coal would not be appropriate.

The thesis that the Clean Air Act and energy policy were in conflict rested heavily on the idea of diminishing natural gas supplies. The curtailments in the early 1970s, and the shortages in the winter of 1976-77 convinced many that natural gas could not be counted on as a fuel for industrial or utility operations. Not only was this view held by the Carter Administration and by many influential members of the Senate and the House, but also by many experts outside government. The Carter Administration's proposals to restrict oil and natural gas use in industrial and utility boilers passed the Congress with little controversy, compared to bitter debate over other parts of the Carter National Energy Plan. So pervasive was the shortage phobia that even decorative uses of natural gas--the fashionable gas lamps used to decorate many neo-colonial houses--were banned.

It is ironic that the 1978 National Energy Act restricted uses of natural gas, particularly for utility boilers, and simultaneously made more natural gas available to gas-starved areas of the country. Prior to passage of that legislation, regulated prices were considerably lower in the interstate market than in the decontrolled intrastate market, the latter representing gas produced and used in the same state. Not surprisingly, most new gas supplies were being diverted to the intrastate market, resulting in curtailments in the Midwest and other nonproducing areas. The Natural Gas Policy Act of 1978 erased the distinctions between the two markets, giving producers new incentives to produce natural gas and to market it in the interstate market. Within a few months, newspapers began describing the new natural gas glut created by NGPA, a situation that, in essence, still exists.

Another part of the 1978 energy package, the Fuel Use Act, was designed to restrict natural gas use. Natural gas was banned from use in new utility and industrial boilers unless the Department of Energy provided an exemption and utility use of natural gas was to be phased out by the 1990s. In essence, federal legislation helped create a natural gas surplus in the interstate market and simultaneously restricted natural gas use. By 1981, Congress virtually eliminated the requirements for utilities to phase out natural gas and the Reagan Administration promised light-handed implementation of the Fuel Use Act. But the Fuel Use Act still stands as a possible restriction to natural gas use.

After much pessimism over natural gas supply through much of the 1970s, current predictions are more positive. Since 1973, reserve additions have been increasing, from a low of one-fourth of production in 1973 to an excess of production in 1981. This surge in reserve additions, however, has occurred mainly from extensions and additions to existing fields, not from new discoveries. Nevertheless, for some years natural gas deliverability will be in excess of demand, as higher natural gas prices spur conservation and some switching to other energy sources. The natural gas market has been transformed, at least for the present time, from being limited by supply, to being limited by demand. The industry's challenge for the future will be to entice customers to stay with natural gas, not to manage curtailments.

Oil is a different story. There are many public policy reasons to be concerned about the level of oil imports, and hence the levels of domestic oil consumption and production. As a major consumer of oil imports, U.S. oil demand will help shape OPEC prices. While this market or monopsony power may not seem as relevant in today's glut, over the long-run, the level of U.S. oil demand will have a strong impact on prices. If the world oil price were higher because of larger U.S. oil imports, then it would add to the cost of all imports. For example, let us assume that some time in the future an increase from 5 to 6 million barrels of oil imports results in raising future world oil prices by $2 per barrel. Not only would the additional 1 mbd of imports result in an additional cost of $2 per barrel for these new imports, but also for the 5 million barrels already being imported. Hence, each additional million barrels adds $2 million per day for the new imports and $10 million per day for the original imports; the additional cost per barrel of oil imports under this example is $12 per barrel.

Put simply, the lower our demand, the lower the prices that OPEC can charge, giving us a continued reason to reduce imports. A study by the Energy Modeling Forum, entitled World Oil, estimates that the hidden costs of importing oil are in the neighborhood of $8 per barrel for the United States and $30 per barrel for the OECD nations. Most of these bidder costs are attributable to the impact of import demand on world oil prices.[1]

The vulnerability to severe economic shocks from oil interruptions is another strong reason to reduce oil imports. The last two price shocks led to double digit inflation, huge economic losses, and unemployment. The chief economist of the OECD estimates that in terms of lost economic growth the 1979 Iranian oil shortage cost the OECD nations $360 billion in 1980 and $620 billion in 1981.[2]

OIL POLICIES

Oil policy should be aimed at reducing these external costs of oil imports through proper incentives for reducing oil consumption and increasing domestic oil production. The federal government has an additional responsibility to protect against oil interruptions. An enlightened oil policy would be based on market intervention, as a natural reaction to both cartel price-setting and the security risks. Government intervention could take the form of taxes and incentives, developing oil stockpiles, and developing technological options; it need not invoke regulation that hampers market operations.

The optimum policy--a term normally saved for policies that are politically infeasible--would be a tariff on imported oil. To reflect the external costs now encountered with importing oil, a tariff of around $10 per barrel or 30 percent of the cost of imports could be justified by the oil premium discussed earlier. By raising the domestic costs of using petroleum products by about 25 cents per gallon, an oil import fee would create a strong incentive for greater fuel efficiency. By increasing the price for domestic crude oil, greater domestic production would be encouraged, and collections from both the fee and revenues from the Windfall Profits Tax would help offset large projected deficits or provide relief from the higher energy prices.

The oil import fee would assure that consumers make the correct economic decisions. For example, it would raise the cost of crude oil by $1.70 per million Btu over coal, a differential almost equal to the cost of coal itself. By forcing consumers to face the real costs of using oil, imposition of a tariff would reduce the need for cumbersome government restrictions on oil use.

A tariff, however, has many political liabilities. Home heating oil users would again pay considerably more for fuel than natural gas users, unless natural gas prices were allowed to gravitate to the same level. United States export industries, such as petrochemicals, will assert that higher petroleum prices will adversely affect their competitive position. U.S. domestic industries will assert that foreign competition has cheaper energy costs. In other words, a tariff has opposition from almost all quarters and very little support from any special interest constituency. Moreover, it could create international competitive problems, and would spark new inflationary pressures.

Since over half of all oil is used in transportation, a large increase in taxes on motor fuels would have similar benefits. Such a tax increase would induce greater conservation without causing the same types of international competitive problems and it would not require concomitant increases in natural gas prices. Most European countries have gasoline taxes of around $1.50 per gallon (3.785 liters). On the other hand, higher gasoline and diesel fuel taxes would not induce greater production, nor would they provide the proper signals to the industrial sector, and they would also spawn more inflation. Despite its limitations, a large motor fuels tax appears to be a more attractive way to reduce oil imports and provide an additional source of government revenue than a tariff on all imported oil products.

Since import tariffs or higher gasoline taxes are both politically unattractive, supplementary actions could be considered. As long as oil costs consumers less than it costs society as a whole, it will be overused--adding to higher total costs and increased vulnerability to interruptions. Government programs that would reduce imports at costs below $10 per barrel, such as tax credits and other forms of subsidy, could be economically justified in a theoretical sense. As a practical matter, however, such subsidies are almost impossible to administer in a fair and efficient way. Continuation of the current additional 10 percent investment tax credit for facilities using solid fuels, as well as disallowance of the credit for oil-fired facilities, would make sense as a modest, partially compensating policy.

NATURAL GAS POLICIES

The natural gas market is in a state of turmoil. While supplies appear abundant and the fuel is clean and convenient, the transition from a complex regulatory structure to a free market will be painful. The road to ultimate decontrol may be pock-marked by prices temporarily overshooting their market clearing price, and some high-cost gas becoming uneconomic as the cushion for rolling in supplies begins to diminish. Long-term take or pay contracts, often indexed to the price of distillate fuel oil, or higher, should cause gas prices to overshoot the price of residual fuel oil, the nearest competitor to natural gas in the large industrial market, unless voluntary negotiations can help drive prices down. Even in the long run, the industry structure--divided between producers, pipelines, and distributors--prevents market signals from being transmitted correctly. And public utility commissions will play an increasingly important role in allocating fixed costs between industrial, residential, and commercial customers.

The fragmented nature of the gas industry and regulation at both state and federal levels, will make it difficult to compete with residual fuel oil. Most of these transitional problems will affect the distribution of economic rents among producers, distributors, and pipelines, and the relative competitiveness between oil and natural gas more than they will affect the competition between coal and natural gas. But higher natural gas prices--whether induced by the Natural Gas Policy Act or new legislation--will have a number of impacts. They should result in improved efficiency of use, enhanced exploration and development of domestic supplies, and greater interfuel competition. Over the long-term, full decontrol of natural gas prices will result in greater efficiency in production and a more stable market.

THE CLEAN AIR ACT

No obvious case exists for drastic surgery to the Clean Air Act. Its impacts on industrial decisions are relatively minor and sufficient gas supplies are available to meet particular constraints in certain regions. With over five years experience under the 1977 amendments, there may well be good reason to streamline the Clean Air Act for other reasons. But these changes must be based on calculations of benefits and costs of alternative policies, not on the need to stimulate greater coal use in the industrial sector.

The Clean Air Act's Prevention of Significant Deterioration (PSD) provisions are particularly controversial. These provisions, which establish tighter emission limits for already clean regions of the country, probably limit massive conversion to coal in certain areas now, such as the Gulf Coast or in complex terrain. But these areas can probably accommodate most of the conversions that will be seriously contemplated.

In order to comply with the Clean Air Act, or state requirements for that matter, it will be necessary under some conditions to use natural gas. But the price increases envisioned under either the NGPA or accelerated decontrol should result in more than ample supplies being available to the industrial market; indeed, the natural gas industry will be fighting hard to retain its current large share of the industrial market. In essence, there are no supply constraints that preclude continued use of gas in heavily developed areas of the Southwest, the Gulf Coast, or in other areas where natural gas is available to stay within the PSD limitations.

In some areas or locations, however, natural gas potential is limited. If environmental limitations precluded coal use in areas where natural gas was not physically available, including any prospective policies to reduce acid rain deposition, then a minor conflict exists between environmental goals and the objective of reducing oil imports. If an oil import tariff were imposed, this relatively limited problem would be partially overcome by forcing consumers to face the true costs of using oil, including the external, hidden costs. The higher costs would create stronger incentives to conserve oil and, on the margin, to convert to coal or other abundant energy resources through use of scrubbers and low sulfur coal.

Changes in the Clean Air Act should be justified on criteria other than energy policy, such as balancing costs and delays against the public health protection sought by environmental requirements. Considering both the minor impact environmental standards currently have in inhibiting coal use, and the ample availability of natural gas supplies to cope with those exceptions, no dramatic change in the Clean Air Act appears necessary.

A NEW INSTITUTIONAL FRAMEWORK

The biggest obstacle to coal use is the institutional structure that forces large, risky, and capital-intensive coal conversion projects to compete for capital allocations with profit-making investments in a firm's main line of business. The current system achieves sub-optimal results, resulting in a smaller amount of coal used in boilers than would be economically determined, a somewhat smaller amount used in process heaters, and a diminished level of cogeneration. The economic distortion is even larger when oil is the alternative fuel. Overall, current institutional arrangements result in a diminished market for coal, despite generally favorable economics.

Manufacturing firms are not generally excited about financing and operating complex and capital-intensive energy systems. Besides all the financing problems discussed earlier, management of a coal-fired facility is complex and labor-intensive. If options exist to avoid financing and operating large energy facilities, many manufacturing firms would be interested in taking advantage of them. The increased use of electricity by industrial firms may at least partially reflect a willingness by industry to trade off higher fuel costs for management convenience and avoidance of capital commitments.

There is no inherent reason why industry needs to provide its own steam energy. Steam is sometimes provided as a utility service, although rarely to industrial operations. Industrial electricity, only a step removed from steam power, is currently provided almost exclusively by electric utilities, but only thirty years ago, industry provided 30 percent of its own needs. Scallop, Inc., a subsidiary of Royal Dutch Shell, currently provides heating and cooling services to commercial customers on a fixed price arrangement. Other firms are attempting to offer various forms of energy services on a fee basis.

Restructuring the way energy services are provided would help overcome many of the impediments to using coal. If coal use could be accomplished without diverting capital from productive investments and without a large management and operational commitment, then coal might compete more closely on the basis of costs. But that would require a dramatic change in the way industrial energy facilities are financed and operated.

To understand how the industrial energy market could be restructured, it is useful to look at four different market functions as they are performed now and how they could be performed with a different institutional structure (see Table 5.1). Today, an industrial firm would handle almost all aspects of a coal conversion, from initial planning and design, to operation. The firm would decide to proceed based on in-house or consultant studies, raising funds from its own cash flow or from debt and equity markets. Once the facility was designed and constructed, the firm would operate the plant with its own employees, usually taking on new staff for this function. It would, of course, consume the energy in manufacturing its product, either as steam or as direct process heat. Finally, the firm may well produce some or all of its electricity; it may even sell the excess to a utility. Needless to say, firms without experience in coal use find the financing and operation functions difficult, and most would find generation of electricity and negotiation with utilities an added and unwanted burden.

If all these functions could be combined under an entity that would coordinate financing, construction, and operation of coal facilities for industrial firms, including, where appropriate, generation of electricity and sale of excess power to utilities, then many of the barriers to coal use would be overcome. Although in a few cases facilities serving a number of plants would be possible, this proposal envisages that coal boilers would be built on or adjacent to a specific firm's plant site. The new entities would plan and supervise construction of industrial coal and cogeneration

TABLE 5.1

NEW FRAMEWORK FOR INDUSTRIAL ENERGY SERVICES

	Investors	Operating Companies	Industrial Firms	Utilities
Role:	Provide investment funds and perhaps entrepreneurial role	Operate facility and perhaps provide entrepreneurial role	Purchase steam under long-term contract	Purchase electricity at its avoided costs
Goal:	Relatively safe investment with opportunity to capture tax benefits	Steady revenue streams of the former; relatively safe investment and tax benefits for latter	Avoid capital and operating costs for provision of steam	Avoid capital costs for new electrical capacity
Benefits:	Good return and tax benefits	Steady operating profit with minimum risk	Avoidance of committing capital; reduced management hassle	Power at no greater cost than if utility provided it; reduction of new capacity needs

projects, operate them at the plant site, and, where appropriate, contract for sale of excess electricity to utilities. Such entities would either provide investment funds directly or would work on a fee basis for other investors. Their revenues would be derived from the sale of steam, and, where appropriate, electricity to industrial firms, and from sales of excess electricity to utilities. But whether involved in the full range of financing and entrepreneurial functions or merely in operations, they would have basic expertise in running industrial coal-fired systems and cogeneration. Such expertise is critical if these entities are to have credibility with industrial customers.

These new energy service companies would act as both investors and operators of energy plants, although the two functions could be separate. The investors could own equity in a company that provided energy services, or could participate in a limited partnership to provide service at one or a few facilities. In the former model, the operating company would be owned by the investors, similar to any other company. In the latter, the investors and the operating company would be separate, with the operating company's profits coming from fees and whatever equity participation it had in the project. The creation of the first model would be quite an undertaking, although the fixed-price energy services provided by Shell's Scallop subsidiary is a partial example of what is envisaged. The limited partnership model is represented by the many project financing arrangements that have become increasingly fashionable.

The consuming firm would purchase steam and possibly electricity at a cheaper rate than self-generation of oil or gas. Although this would be an attractive proposition, many firms will be concerned that dependable supplies may not be available. While some assurance can be provided in the contract between the industrial firm and the energy service company, the reputation of the energy service company will be extremely important. Most firms will probably keep their oil or natural gas facilities as a backup, in case the energy service company cannot deliver for some reason. Actually, by having the capacity to burn many fuels, the industrial firm should improve its energy security.

To protect the energy service company, on the other hand, the manufacturing firm would need to be willing to enter into long-term contracts, with a penalty for early withdrawal from the system. Some firms will worry about this limit on their operating flexibility.

The utility would be a third party to these arrangements if cogenerated power were sold into the grid. Regulations implementing the Public Utilities

Regulatory Policies Act (PURPA) require utilities to pay cogenerators prices equivalent to the costs that utilities avoid from burning expensive oil or adding new capacity. Many utilities find the purchase of cogenerated power advantageous if they are attempting to avoid adding new capacity, and most negotiate prices at less than their avoided costs in exchange for longer-term contractual arrangements. Indeed, one could foresee utilities being interested in establishing energy service companies as subsidiaries themselves. Unlike some diversification options pursued by utilities, operating coal-fired cogeneration systems would build on many utilities' strength in operating power plants.

Current federal law limits utility ownership to less than 50 percent of a cogeneration facility if the utility is to take advantage of the tax benefits and the avoided cost level of payments. This limitation has discouraged many utilities from becoming active in cogeneration projects. Although difficult to predict, it is likely that utility willingness to become involved with cogeneration projects would increase substantially if the current restriction on utility ownership were removed.

New institutional arrangements are only likely to evolve if each of the major actors foresees strong financial and other benefits. The industrial firms should be interested in cheaper steam and electricity without tying up capital or entering into long-term commitments. Investors would like reasonable rates of return, an opportunity to take advantage of tax and regulatory benefits, and relatively low risk. Operating companies might be interested in either equity positions in coal cogeneration projects or in relatively low-risk fee operating contracts. Many utilities should be interested in purchasing cogenerated power as a way to reduce future capital requirements, and, in some cases, they may choose energy service companies as a source of diversification investment.

Current federal policy favors energy service companies that provide coal and coal cogeneration energy services. Investment tax credits (ITC), accelerated write-offs, and PURPA incentives represent powerful economic inducements for these types of projects. The full benefits of energy service corporations are just coming into public view. It was not until 1980 that coal became substantially more attractive than oil. PURPA regulations, requiring utilities to pay the avoided costs for electricity purchases from cogenerators and small power producers, were not promulgated until March, 1979. The rather generous tax benefits from the Economic Recovery Tax Act were not available until 1981. By the time the full panoply of incentives was available, the United

States was in the grips of a serious economic downturn and interest rates were soaring, creating a decided reduction in business investment. Hence, as yet, the potential market for energy service companies has not really been tested, although various entrepreneurs have apparently signed up some customers.

The future for creating many new entities that provide energy services is far from certain. Although conceptually this form of energy service company is not much different from companies that lease or provide other services on a contractual basis, there are few actual cases where third parties have ever provided energy services. The investment requirements for large coal-fired boilers are huge, often in the hundreds of millions of dollars. Hence, it is likely that initial forays into these institutional arrangements may require companies with established reputations and substantial resources, such as major engineering and construction firms, boiler manufacturers, or electric utilities. Considering current uncertainties about world oil prices, creation of these new entities may be slow. It may take some combination of rising economic activity, and a prognosis for higher oil and gas prices to stimulate an energy service industry.

CONCLUSION

Many will find this a curious place to end this report. When this study began, it seemed reasonable to assume that environmental requirements were inhibiting coal use, at least to some extent. But the more the industrial decisionmaking process was understood, the more obvious the conclusion that environmental standards, although almost uniformly considered a nuisance by industry, played a very small role in actual coal-use decisions. Coal conversions are generally not made because they represent a large diversion of capital away from productive investments and do not materially affect a firm's competitive position. If industry had the option to purchase steam directly--without commiting its own capital and management effort--then a major obstacle to coal use could be overcome.

President Kennedy once remarked that myths were much more dangerous than lies. The latter could almost always be disproved at some point, but myths, by containing a kernel of truth, tend to linger. The commonly repeated refrain that environmental requirements have played a large role in discouraging industrial use of coal is such a myth. The contrary conventional wisdom that coal-use decisions are strictly economic and that coal use will shoot up in the future represents a similar myth. These two myths cloud public policy to some extent, and prevent us from understanding how industrial decisions are made and what the likely fuel-use mixes are for the future. Only when these myths are thoroughly debunked can we improve our understanding of the future role that coal will play in the industrial market.

NOTES

1. Energy Modeling Forum, World Oil: Summary Report, February 1982, p. 5.

2. Sylvia Ostry, "A View from the OECD World" (Speech delivered before the Atlantic Institute for International Affairs Conference, October 22-24, 1981), p. 6.

APPENDIX A

Major Air Quality Programs Affecting Industrial Fuel Choice

The regulatory system set up under the Clean Air Act of 1970 contained two major breakthroughs in the control of air quality. First, the act required the EPA to establish uniform national air quality standards--National Ambient Air Quality Standards (NAAQS). Each state was to develop programs to meet the NAAQS, called State Implementation Plans (SIPs). EPA was also required to establish uniform industrial emission standards in the form of New Source Performance Standards (NSPS) for new stationary sources of air pollution and National Emission Standards for Hazardous Air Pollutants (NESHAPs), for both new and existing sources of toxic air pollutants.

These programs, explicitly detailed in the act, were strengthened by the addition of two other programs, the Emissions Offset Policy (EOP) and the Prevention of Significant Deterioration (PSD) programs. The PSD and EOP programs were not set forth in the 1970 act itself, but resulted from EPA and judicial interpretations of the act. While some critics objected to EPA's development of these programs without precise statutory directives, these programs were codified, and in some cases strengthened in the 1977 Amendments to the Clean Air Act.

NATIONAL AMBIENT AIR QUALITY STANDARDS

NAAQS have been established for the following pollutants: sulfur dioxide (SO_2), particulate matter (PM), nitrogen dioxide (NO_2), carbon monoxide (CO), ozone, and lead.[1] The standards are the level of air pollutant concentration deemed acceptable to protect public health (primary standard), and welfare (secondary standard). The 1970 act required that NAAQS be attained by 1976. When it became evident that some areas would not attain the NAAQS, the 1977 amendments required every state to determine how the air quality measured up to the national standard. These determinations were expressed in geographical units depending on the air quality monitoring data available. Three determinations could be made for each area: non-attainment (air quality worse than the standards), attainment (air quality better than or equal to the standards), or unclassified (data not available).

Under the 1977 Amendments to the Clean Air Act, those areas of the country not meeting the standards were required to meet the standards by December 31, 1982 (or under certain conditions, December 31, 1987, for CO and ozone). Efforts to attain the NAAQS will restrict the amount and the type of new pollutant emissions allowed. Since the areas of the country not meeting the NAAQS, particularly for SO_2 and PM, are generally industrial areas, attainment of the NAAQS has the potential to limit industrial expansion. Because coal combustion emits major quantities of SO_2 and PM, in addition to lesser amounts of CO, NO_2 and trace elements, NAAQS attainment is an important consideration in the fuel use decisionmaking process.

AIR POLLUTION CONTROL TECHNOLOGY

The 1970 Clean Air Act requires EPA to set a national emissions standard for certain categories of new stationary sources, one of which was fossil-fuel-fired steam generators. This minimum national standard was designed to prevent states from being unduly lenient toward sources of air pollution in order to attract industry at the expense of public health.

In December 1971, the EPA set emission limits for SO_2, PM, and NO_2 from fossil-fuel-fired steam electric generators with a heat input greater than 250 mmBtu/hr burning solid or liquid fuel.[2] Boilers this size are generally utility boilers or extremely large industrial boilers. The 1971 standards allowed low sulfur fuel to be burned to meet the standard, and did not require a technological pollution control system such as a flue gas desulfurization system (scrubber).

Since that time, several technological advances occurred that indicated the fossil-fuel-fired steam generator standard should be reexamined. As a result, the 1977 Clean Air Act Amendments required that the NSPS be studied.[3] The 1977 amendments also appeared to require that the NSPS provide for a percentage reduction standard in addition to an emission limit. In 1979, EPA promulgated a new set of utility standards that required some use of scrubbers for even the lowest sulfur fuels.[4]

In addition, for the first time, the federal government was to regulate industrial boilers below 250 mmBtu. The EPA is still in the process of determining the emissions limitations for industrial boilers; at some point, industrial boilers could be required to install scrubbers to meet the pollution control requirements.[5]

Two other technological control standards govern the pollution control equipment to be installed on industrial boilers. First, the Prevention of Significant Deterioration (PSD) program requires that new sources subject to PSD rules install the Best Available Control Technology (BACT) for each pollutant emitted from the facility. Naturally, coal-fired industrial boilers are quite likely to be subject to these rules if the boiler is large enough to be regulated under the PSD program. BACT was intended to be more stringent than the NSPS; in practice, NSPS and BACT are generally identical for many categories of sources.[6] Second, sources locating in a non-attainment area must achieve the Lowest Achievable Emission Rate (LAER) that is feasible. This also represents a significant technical constraint on industrial boiler emissions, which is supposedly more stringent than NSPS.[7]

AIR QUALITY AND THE STATES

Superimposed on the federal scheme is the air quality program of each state. Besides the SIPs that are required under federal law, most states have their own set of regulations. In some states, there are also county and local air quality regulations. This pastiche of rules and regulations--federal, some state, some local--must be met by industry.

EMISSIONS OFFSET POLICY--NON-ATTAINMENT AREAS

By 1975, when it became obvious that the air quality standards established after the 1970 Clean Air Act would not be met, the EPA began to explore solutions to meeting the standards while maintaining economic growth. The result was the EPA Emissions Offset Policy, released on December 21, 1976. This policy required firms desiring to locate or expand operations in a non-attainment area to meet several requirements. Besides meeting strict technological control requirements, the plant had to obtain pollution offsets greater than the expected pollution from the plant. It either had to reduce emissions at its own facility or obtain offsets elsewhere, leading to what EPA termed a "net air quality benefit."

While the Emissions Offset Policy encountered substantial opposition, Congress utilized it in the 1977 amendments as a means to reconcile growth with improvement of air quality in non-attainment areas. After July 1, 1979, no growth was to be permitted unless a revised SIP approved by the EPA demonstrated attainment of the primary NAAQS by December 31, 1982 (with an extension until December 31, 1987, for hydrocarbons (HC) and CO under certain circumstances). During the period between 1979 and 1983, states could adopt one or both of two strategies to accommodate growth in non-attainment areas. First, they could adopt an offset policy which assured that total emissions of existing sources, new minor sources, and permitted new major sources would be less than the total emissions at the time of the new source permit. Second, they could tighten controls on existing sources more than the amount necessary to reach attainment, thus creating a growth allowance. Either strategy satisfies the "reasonable further progress" requirement of the 1977 amendments.[8]

PREVENTION OF SIGNIFICANT DETERIORATION

Prompted by a Supreme Court decision preventing significant deterioration of air quality in any clean region of the country, EPA responded by establishing the Prevention of Significant Deterioration (PSD) program.[9] That program requires major new or modified pollution sources in attainment or unclassified areas to conduct analysis and install pollution control technology to ensure that air quality would not be significantly impaired. The PSD regulations were issued in December of 1974, and although EPA was immediately taken to court, the regulations were operative from that date.[10] Congress, in the 1977 amendments, adopted the entire PSD program as conceived by EPA, but tightened up the allowable emission levels.

The PSD program, through the permit and analysis requirements, attempts to limit new air pollution concentrations to a uniform amount over current pollutant concentrations. Attainment areas are subdivided into three classes, and maximum allowable increases (increments) in SO_2 and total suspended particulates (TSP), are specified for each class. These increments are the amount of pollution allowed over current pollution levels. A baseline level of pollution must be determined from which increments are allowed, requiring industrial firms to do the monitoring in areas where data is not available.

The 1977 amendments designated all large national parks, wilderness areas, and other unique natural attractions as Class I areas, where stringent increments allow almost no growth. All other areas of the country are designated as Class II, where moderate increases in pollution levels are allowed. States can redesignate Class II areas as Class III areas, where increases in pollution levels are allowed up to roughly 50 percent of the national standard. However, no state has yet redesignated an area as Class III. States can also redesignate Class II areas as Class I, which has occurred in at least one instance. The PSD program held increased emissions in Class I areas to roughly 2.5 percent of the annual ambient air quality standard, Class II areas to roughly 25 percent and Class III areas to 50 percent.

The 1977 amendments established a preconstruction review process designed to determine whether the source will violate the PSD increment or the NAAQS. Most industrial boilers emit enough pollutants to be included in this review process. Air quality modeling and monitoring are usually necessary. Again, since coal emits SO_2 and PM, coal-fired industrial boilers will be limited either by the PSD increments or the NAAQS.

Aspects of the PSD regulations promulgated by EPA pursuant to the 1977 amendments were challenged by industry in the D.C. Circuit.[11] That resulted in promulgation of new PSD regulations,[12] which have in turn been challenged.[13]

NOTES

1. 40 C.F.R. §§50.4-.12 (1982). The NAAQS have been the subject of considerable controversy concerning their scientific validity. The D.C. Circuit struck down issuance of a secondary SO_2 standard as being unsupported by substantial evidence. Kennecott Copper Corp. v. EPA, 462 F.2d 846 (D.C. Cir. 1978). The controversy concerning the scientific basis for the NAAQS led Congress to mandate periodic review of the standards. EPA is currently in the process of reviewing the CO, SO_2, NO_x and PM standards. 48 Fed. Reg. 18474 (April 25, 1983). EPA revoked the primary and secondary hydrocarbon (HC) standard because of the uncertainty about the relationships between ambient hydrocarbon concentrations and photochemical oxidants, with respect to human health and welfare. 48 Fed. Reg. 628 (January 5, 1983).

2. This new source performance standard (NSPS) is at 40 C.F.R. §60.40-.46 (1982). It sets a particulate matter emission limit of .10 lb/mmBtu [.0454 kilogram (kg)/mmBtu], as well as capacity limits for all fossil fuel and wood boilers. The sulfur dioxide emission limits are .8 lb/mmBtu (.3632 kg/mmBtu) for oil and oil-wood boilers, and 1.2 lb/mmBtu (.5448 kg/mmBtu) for coal and coal-wood boilers. The NO_x emission limits are .2 lb/mmBtu (.0908 kg/mmBtu) for natural gas and gas-wood boilers, .3 lb/mmBtu (.1362 kg/mmBtu) for oil and oil-wood boilers, and .7 lb/mmBtu (.3178 kg/mmBtu) for coal and coal-wood boilers.

3. 42 U.S.C. §7411(b)(1)(B) (Supp. II 1978).

4. 40 C.F.R. §60.40a-49a (1982). The standard imposes an emission limit of .03 lb/mmBtu (.01362 kg/mmBtu) of particulate matter for all utility boilers greater than 250 mmBtu/hr as well as a percent reduction in particulate matter of 99 percent for coal and wood-fired boilers and 70 percent for oil-fired boilers.

The sulfur dioxide limit for coal-fired boilers is 1.2 lb/mmBtu (.5448 kg/mmBtu) and a 90 percent reduction, or .6 lb/mmBtu (.2724 kg/mmBtu) and 70 percent reduction. The sulfur dioxide limit for oil and gas-fired boilers is .8 lb/mmBtu (.3632 kg/mmBtu) and 90 percent reduction, or .2 lb/mmBtu (.0908 kg/mmBtu) with no percentage reduction. The NO_x emission limit is .5 to .6 lb/mmBtu (.227 kg to .2724 kg/mmBtu) for boilers fired with coal or coal-derived fuels, .3 (.1362 kg/mmBtu) for oil-fired boilers, and .2 (.0908 kg/mmBtu) for gas-fired boilers. The NO_x percentage reduction is 25 percent for gaseous fuels, 30 percent for liquid fuels, and 65 percent for solid fuels.

5. An advance notice of proposed rulemaking was published on June 28, 1979. 44 Fed. Reg. 37632. An initial draft developed during the Carter Administration NSPS required a 50 percent SO_2 reduction in boilers below 250 mmBtu/hr and no percent reduction (i.e., no scrubbing) in boilers below 150 mmBtu/hr.

The most recent draft proposal would set emission limits for boilers greater than 29 mw (100 mmBtu/hr) heat input. The particulate matter limit would be .1 lb/mmBtu (.0454 kg/mmBtu) for wood, coal-wood, and coal-fired boilers with wet scrubbers. No particulate limit was proposed for oil-fired boilers. The NO_x limit would be .7 lb/mmBtu (.3178 kg/mmBtu) for pulverized coal-fired boilers, .6 lb/mmBtu (.2724 kg/mmBtu) for other coal-fired boilers, and .2 lb/mmBtu (.0908 kg/mmBtu) for gas or oil-fired boilers. The draft did not include SO_2 limits on the basis of the percentage reduction requirement.

6. National Commission on Air Quality, _To Breathe Clean Air_ (Washington, D.C., March 1981), pp. 166-167.

7. National Commission on Air Quality, pp. 138-139.

8. 42 U.S.C. §§7401-7642 (Supp. II 1978).

9. _Sierra Club_ v. _Ruckelshaus_, 344 F. Supp. 253 (D.D.C.), _aff'd per curiam_ 4 Envir. Rep. (BNA) (Envir. Rep Case.) 1815 (D.C. Cir. 1972), _aff'd by an equally divided court sub. nom._, _Fri_ v. _Sierra Club_, 412 U.S. 541 (1973).

10. 39 Fed. Reg. 42510 (1974).

11. _Alabama Power Co_. v. _Costle_, 636 F.2d 323 (D.C. Cir. 1979).

12. 45 Red. Reg. 52576 (August 7, 1980).

13. _Chemical Manufacturers Assn_. v. _EPA_, No. 79-1112 and consolidated cases (D.C. Circuit).

APPENDIX B
Case Study Methodology and Summaries

The basic research for this study consisted of the following:

- o an analysis of fuel-use trends in eight large energy-using industries, including an in-depth case study analysis of one fuel conversion decision in a particular firm in each of the eight industries;
- o two case studies of small firms;
- o over thirty interviews with other industrial firms.

Each case study description begins with a general characterization of the industry--its energy-use patterns, fuel-use costs, fuel supply experience, barriers to coal use, and environmental regulatory experience. They then move to a discussion of a particular firm and that firm's response to one fuel-use decision at a particular facility. The case studies conclude with a look at the corporate decisionmaking process.

By focusing on discrete fuel-use decisions at one particular firm, the report was an attempt to understand the relative importance of economic, behavioral, and environmental factors in the final decision. In addition, anywhere from two to six more firms were interviewed by telephone to round out our understanding of the industry and to assure that the experience of the firm reviewed was not unique. Because smaller firms would be more likely to have problems with environmental requirements, we evaluated one medium-sized brick company and two small companies--a small specialty paper manufacturer and a rose grower.

Since fuel-use patterns differ regionally due to the availability of fuel transportation networks, some attempt was made to obtain a geographic spread in the actual fuel-use decisions considered. Wherever a firm had experienced multiple fuel-use conversion decisions, we attempted to select a location that filled a geographical gap. The actual geographic spread was as follows:

Gulf states	4
Southeast	1
Northeast	3
Mid-Atlantic	1
Midwest	1

The bias towards the Gulf states is attributable to the rapid industrial growth of this area. Two out of three Northeast cases involved small firms.

The study covered most of the major energy using industries in the United States. Specifically, it included detailed analyses of firms in the aluminum, cement, chemicals, food, pulp and paper, petroleum refinery, and steel industries. The case studies also focused on the brick industry and on a small rose grower.

The analysis for Chapter 4 was augmented by interviews with an additional ten firms. These interviews focused on the firms' experiences with coal conversion, the environmental problems they encountered, and the strategies they would use to overcome environmental restrictions.

Overall, the study was able to capture a good cross-section of industry fuel-use practices. The results appear robust considering the consistency of so many of the major findings. The combination of case studies and detailed interviews appears highly preferable to the use of questionnaires or other methodological techniques. The information our methodology revealed underscored the need to understand how and why certain decisions are made before concluding that any one facet of public policy is the dominant force motivating industry behavior.

THE ALUMINUM INDUSTRY: Kaiser Aluminum and Chemical Corporation

I. INDUSTRY PROFILE

A. Function

The aluminum industry provides a product of increasing importance in an energy-conscious world: a lightweight, strong building or fabricating material made from an abundant mineral. Aluminum mills produce primary aluminum in several basic shapes (rods, foil, sheet, etc.), which are subsequently transformed into the more familiar consumer products (siding, cans, vehicle components, etc.) by secondary fabricators. Secondary fabrication is not considered in this study because of its relatively low energy use compared to primary production.

Primary aluminum production entails four steps: mining of ores, refining ores into alumina, electrolytic reduction of alumina into aluminum, and casting of molten metals into basic mill products. The latter three stages are frequently integrated at single or adjacent plants, and together account for almost all energy used in aluminum production.

B. Energy Use

Each of the three stages uses energy in slightly different forms. Alumina refining uses steam from boilers to heat digesters for separating the aluminum oxide from the raw ore and direct heat in calcinating kilns to fire off excess moisture and produce pure alumina powder. Electrolytic reduction of alumina (or smelting) in cells employs only electricity to produce aluminum; this stage is the most energy intensive, consuming 70 percent of all energy used in aluminum production. Finally, casting uses both direct and indirect heating of the metal in treating and holding furnaces.

Electricity represents 65 percent of all energy used and less than one-quarter of this electricity is self-generated. Because of the location of primary aluminum producers, a good deal of the electricity is generated by natural gas or hydroelectric facilities. Since 1972, coal- and nuclear-fired electrical generation have begun to displace some of the natural gas-fired generation. Natural gas still provides 25 percent of the energy used in aluminum production.

Although aluminum producers have reduced total energy demand by 10 percent since 1972, promising opportunities remain. The average smelter, for example, uses 8 kwh/lb. [8 kwh/.454 kilograms (kg)] of product--considerably more than the 6.1 kwh/lb. (6.1 kwh/.454 kg) required when state of the art technology is used.

C. Economics

Because of massive economies of scale, the aluminum industry is highly concentrated with only three major domestic companies. These firms compete on the basis of price and, therefore, costs. Rapidly increasing energy costs have put pressure on the industry to modernize plants and equipment to take advantage of new processes.

Unfortunately, the aluminum producers are in the same financial position as most other basic industries. Poor market performance during recent recessionary years has limited the financial resources for necessary investments at the industry's disposal. Fuel switching projects must compete with other pressing demands on capital.

II. FUEL-USE PARAMETERS

A. Technical Barriers to Coal

With 65 percent of energy use supplied by electricity and most of that purchased from utilities, the greatest impediment to coal use in aluminum production lies in the utility sector. Self-generation is possible, particularly at integrated mills where cogeneration is economic. But most direct coal use will come in using coal for the alumina digester boilers and for holding furnaces.

Several steps of the production process cannot be coal-fired. Calcinating kilns and direct heat treating furnaces require clean fuels because of product contamination concerns. Pre-baking of the anodes used in electrolytic reduction is performed in small, precise temperature furnaces, for which coal, with its uneven heat characteristics, is inappropriate. Finally, indirect heat treating furnaces face ash buildup problems.

B. Fuel Cost and Supply Experience

Changing natural gas contract conditions in the Southeast and electricity contracts in the Northwest have raised serious concerns about future fuel costs among aluminum producers. Natural gas supply curtailments of the mid-1970s led some producers to switch to oil in calcination and holding furnaces.

C. Environmental Regulatory Experience

Aluminum producers have had to comply with a range of environmental regulations. These requirements have caused a drain on general capital availability for other projects.

III. ILLUSTRATIVE FUEL-USE DECISION

A. Background

Three Louisiana facilities of Kaiser Aluminum and Chemical Corporation (an alumina-chemicals plant, an alumina plant, and an aluminum smelter) using natural gas and located in the same area were considered for fuel conversion. The company initiated the conversion study in 1975, in part, in anticipation of expiring natural gas contracts in 1979, 1981, and 1985 respectively.

B. Options Considered

The two alumina facilities faced a range of options available to produce the steam and electricity required: conversion to oil, self-supplied natural gas, coal gasification, and conversion to coal-fired boilers. In addition, the alumina plant could use a combination of coal conversion and purchased electricity. The smelting facility must use electricity and, therefore, had only three options: conversion to coal-fired, self-generation on site; off-site, coal-fired self generation in a joint venture with another company or utility; or purchased electricity.

C. Major Decision Factors

Kaiser dropped the multitude of non-coal options for the alumina plants for a variety of reasons. Oil was rejected subsequent to dramatic changes in the oil price during the late 1970s. The firm realized that use of self-supplied natural gas would entail severe opportunity costs considering the market value of natural gas. Coal gasification remains, in the company's opinion, unproven in commercial applications. Coal-fired cogeneration emerged as the preferred solution to be pursued sequentially at the two plants, although the first of these investments is still in the planning stages.

The decision about meeting the energy requirements of the smelter has been postponed. The firm is still actively studying its ultimate options, however. Conversion to coal is hampered by spatial constraints because of coal storage and fly ash disposal requirements and the potential objections of neighbors. A joint venture, coal-fired self-generating plant would overcome these obstacles, but it would face the well known institutional problems of such common undertakings. Finally, purchased power offers the least control over cost and availability.

D. Decisionmaking Process

Because of the extensive need for and high cost of modernizations, fuel switching at aluminum facilities (particularly smelters) is decided largely on strategic considerations. The company's competent in-house energy staff and extensive use of special consultants has helped the decisionmaking process become more sophisticated. Kaiser uses a variety of sources for fuel price projections and generally considers a range of natural gas decontrol scenarios. The fuel-use decisions have been reviewed extensively by the top management in the corporation and highlighted in its annual reports.

THE BRICK INDUSTRY: General Shale Products Corporation

I. INDUSTRY PROFILE

A. Function

The production of bricks involves the mining of clay or shale and the preparation of materials cut into brick sizes. The bricks are stacked onto flatbed cars, passed through a drying room to remove excess moisture, and then fired in a kiln. Periodic or beehive kilns for firing had been the accepted method until the advent of tunnel kilns, which combine the drying and firing stages within a single structure. Periodic kilns concentrate the heat source within the kiln and are well suited for burning large quantities of coal. Because a tunnel kiln may have over 200 separate burners along the length of the kiln to regulate temperatures, natural gas has been the preferred fuel for firing the tunnel kiln process.

B. Energy Use

Originally, coal supplied most of the fuel for the brick industry, but price and convenience made natural gas the preferred fuel after World War II. By 1972, natural gas supplied 80 percent of the fuel mix for brick manufacturing. Ninety percent of the energy requirement in brick manufacturing is for direct direct process heat. Since product contamination is not a major problem for the industry, solid fuels offer a feasible alternative to oil and natural gas. Some firms have experimented with wood chips or sawdust.

C. Economics

The health of the brick and structural clay tile industry (Standard Industrial Classification 3251) depends on trends in the housing market, which are sensitive, in turn, to interest rates. The low number of housing starts in recent years resulted in brick production cutbacks, inventory accumulation, and lower profits. Investment capital is, therefore, scarce for many companies. Brick firms are typically small and, though most of them control their own source of the key raw material, shale or clay, they cannot integrate backwards into energy supply with the exception of those firms now using wood. Size of the firm and lack of capital hamper the industry in equipping plants with tunnel kilns, the major change in technology affecting the production of bricks.

Capital costs for installing a tunnel kiln are often beyond the means of small firms where the investment may exceed the total value of the firm. Economies of scale clearly exist for the brick industry, but most firms find themselves unable to take advantage of these efficiencies because of uncertain markets and the high costs of investment. Increased mechanization and a reduction in labor have been characteristic during the last decade.

Almost 200 firms produce brick domestically, but the top 3 firms account for 25 percent of sales. High capital requirements have been the reason for recent consolidation and mergers. The industry is concentrated in the South and Southwest, the areas with the greatest number of housing starts.

II. FUEL-USE PARAMETERS

A. Technical Barriers to Coal Use

The major technological innovation in brick production, the tunnel kiln, originally posed the greatest obstacle to coal use. Because the tunnel kiln requires multiple fuel feed points with precisely regulated temperatures, coal combustion technology did not appear suitable. The company interviewed, General Shale, tackled this problem successfully by developing an engineering solution. In-house research developed a patented pneumatic system to propel granulated coal through numerous small injector lances. The new system allows a controlled fuel feed.

B. Fuel Costs and Supply Experience

Since energy costs are a significant factor in production (13 to 15 percent of total costs), the industry has a continual interest in conservation. From 1972 to 1977, energy input per unit shipment decreased 24 percent largely due to improvements in kiln construction and the introduction of tunnel kilns.

In the mid-1970s, brick manufacturers were hard hit by disruptions in the supply of natural gas. The industry has tried to cope with the situation by keeping supplies of oil or propane on hand and by drawing on brick inventories.

C. Environmental Regulatory Experience

The most significant environmental regulation of the brick industry has been in the area of mining. Manufacturers argue that they should be exempt from the provisions of the Surface Mining Act since shale pits do not produce the tailings, overburden, or acid runoff that were the concerns of that legislation. Because of extensive use of natural gas, the industry has had little experience with air pollution control requirements.

III. ILLUSTRATIVE FUEL-USE DECISION

A. Background

General Shale Products Corporation, an industry leader located in the Southeast, produces a number of construction products. During the last decade, the firm earned higher profits than other companies in the industry; its overall economic strength allowed the company to borrow at the prime rate. The firm does have a debt ceiling, however, and has had to postpone at least one of its conversion projects in order to stay within its capital budget. The case study examined a conversion project at one of the firm's oldest manufacturing facilities.

B. Options Considered

Natural gas supply conditions during the 1970s led the firm to adopt an aggressive fuel switching policy. Coal was the only option considered.

C. Major Decision Factors

The motivating factor for General Shale's fuel switching policy and plant conversions has been concern over the availability of natural gas. As early as 1971, the firm experienced natural gas supply curtailments and began researching the use of coal. At that time, the company depended on oil and natural gas for 90 percent of its fuel needs. The firm is now implementing a corporate policy to convert all plants to coal. In 1979, 65 percent of company fuel use was supplied by coal.

The economics of coal use also recommended conversion because of a unique efficiency in the burning of coal in kilns. Heat input requirements per unit of brick declined almost 30 percent with the use of coal. Natural gas reaches its highest temperature at the burner tip at the injection point of the kiln. When solid coal is propelled into the kiln, the lighter particles burn in suspension and the coarser particles combust more slowly, often in direct contact with the bricks. This produces heat where it is most needed. The added bonus in heat efficiency has resulted in a cost savings of $.005 per brick. The payback period for the total investment is estimated to run between two to five years depending on whether gas or oil is in current use.

Pollution control was an important consideration for reasons other than regulation. The coal-firing process requires a high quality--near-metallurgical grade--coal of not more than 1 percent sulfur and 5 to 7 percent ash content. A higher ash coal requires periodic cleaning of the kiln and results in reduced thermal efficiency. The ash collects in the kiln and acts as an insulator. A low sulfur content is necessary for meeting environmental standards and for controlling the color of bricks. Ironically, the motive for the fuel switch, security of supply, is not totally resolved by the firm's decision. The high quality coal needed for the process already commands a premium price and is in demand for some nonfuel uses; its low "coke-button" makes it unfit for most metallurgical purposes, however. General Shale is confident about its ability to purchase an adequate supply since it is near producing areas.

Since technological necessity required the firm to use low sulfur coal, particulate control remained the firm's only other major environmental compliance problem. The plant is located in a non-attainment area. Most of the environmental permitting process is directed by the state air pollution control agency through district offices. Local regulators have a good relationship with industry and view the use of coal, when environmental considerations allow it, as a benefit to the area. General Shale was able to demonstrate adequate particulate control and the availability of a pollution offset through the use of a baghouse, the retirement of the existing plant, the use of low ash coal, and the pneumatic feeding system developed by their engineers, which forces air into the system instead of blowing coal dust out. A multi-clone was considered as an alternative to the baghouse but was rejected because of poor performance characteristics.

D. Decisionmaking Process

The brick manufacturer, although large for the industry, is a relatively small firm (annual sales of $60 million) with a centralized decisionmaking process. The development and success of the total conversion policy is attributable largely to the diligence and foresight of the company's chief executive, who had been searching for a secure fuel supply source since the early 1970s.

The firm is anxious to see the brick industry adopt the new coal technology since it regards its own strength as dependent on the ability of brick manufacturers to compete with other building material suppliers. It works closely with other brick firms to share information and to pursue conversion to coal.

THE CEMENT INDUSTRY: Ideal Cement*

I. INDUSTRY PROFILE

A. Function

The cement industry produces three varieties of its product (Portland, masonry, and white) by a three stage process. First, limestone is ground and mixed with other materials (silica, aluminum, and iron) to achieve the desired chemical composition. Then, the material is burned in a kiln to produce the chemical transformation into "clinker." Finally, clinker is finish-ground with gypsum. Cement is used principally as a bonding agent in concrete and for block and brick construction.

B. Energy Use

The cement industry turned away from coal use during the 1950s for the same reasons most other industries did (cheap, available natural gas and oil). In 1947, coal provided 67 percent of total energy use in cement manufacture; by the early 1970s, this figure had declined to 35 percent. The post-embargo era has seen a rapid return to coal's ascendancy; in 1979, coal again provided 65 percent of total energy requirements. Although the cement industry has made significant improvements in energy efficiency [from 6.75 mmBtu/ton (6.75 mmBtu/.907 metric tons) to 6.12 mmBtu/ton (6.12 mmBtu/.907 metric tons) in the period 1972-79], a great deal of room remains for additional improvement since the newest kilns use only 3 to 3.5 mmBtu/ton (3.5 mmBtu/.907 metric tons). The bulk of improvement has come from increased use of new kilns, although a limited amount of retrofitting is possible.

---

* At the time the case study was conducted, the company's name was Ideal Cement, but since then it has been changed to Ideal Basic Industries.

C. Economics

The cement industry includes fifty-five firms and concentration is low. The industry operates 168 plants throughout the country close to demand centers because of the high cost of transporting the end-product. Since most cement is used in construction, demand tends to be somewhat cyclical, and annual production is currently well below peak production in 1972. Despite a steady increase in the per ton price, gross margin has declined from 30 to 25 percent, largely due to energy cost increases. A lack of capital has not inhibited coal conversion in this industry.

II. FUEL-USE PARAMETERS

A. Technical Barriers to Coal

Most of the energy consumed in cement production fires the kilns for direct heat processing of the raw materials. As suggested by the early industry fuel mix, coal is compatible with cement production; most of the impurities tend to be absorbed into the clinker, a fact that also reduces sulfur control requirements. The only potential barrier to coal conversion lies in the older plants where the expenditures for additional particulate control equipment may not be worth the investment.

B. Fuel Costs and Supply Experience

The cement industry had been shielded from the full impact of energy price increases because of the extensive use of natural gas. However, gas supply curtailments affected cement plant operations in the early 1970s and served as a major fuel switching impetus.

C. Environmental Regulatory Experience

The cement industry has a long standing familiarity with pollution control requirements and has been using electrostatic precipitators (ESPs) since the 1930s to control kiln emissions. Many of these systems need upgrading to handle the added particulate emissions from coal burning. Many cement manufacturers now use more efficient baghouses although this technological shift reduces the amount of dust available for recycling. Solid waste disposal is a familiar, but expensive, factor of cement production, depending on the proximity of a suitable dust disposal site.

III. ILLUSTRATIVE FUEL-USE DECISION

A. Background

Ideal Cement moved aggressively to coal during the 1970s. The particular fuel switch described below occurred at one of the firm's largest plants, located adjacent to the Houston shipping canal. Prior to conversion, the kilns burned natural gas; particulate emissions were controlled by ESPs.

B. Options Considered

Ideal examined the feasibility and cost of converting the plant to either coal or petroleum coke. Coke was considered at this one plant because of the proximity of major refineries (and, therefore, of inexpensive coke supplies). Residual fuel oil was not considered because the decision was made in 1974, right after the Arab oil embargo.

C. Major Decision Factors

The decision in this case involved a choice between coal or coke. Because the firm had already made a strategic decision to switch from natural gas, the question of the financial worthiness of the project was not important; indeed, the company committed capital to this conversion despite a projected rate of return below the hurdle rate. The total capital requirements for switching from natural gas use at all eight targeted plants equaled less than 8 percent of the total capital cost of a major new facility.

The principal factors considered in opting for petroleum coke over coal were twofold. First, the company ascertained that coke would be plentiful and relatively cheap because of the plant's location. Second, the plant already had ESPs in place and determined that coke, with its lower ash content, would place less strain on the equipment. Of much more minor importance was the firm's concern about spontaneous combustion in a coal pile. Right after the fuel switch, the plant burned a combination of petroleum coke and natural gas (to boost volatility); the gas has since been supplanted with coal.

D. Decisionmaking Process

At the time of this study, Ideal did not have a rigid capital allocation decision flow. Projects could be proposed from any one of a number of loci. Divisional management in charge of a proposal convened in-house managers with special expertise to review the project's merits. If the project appeared worthy of funding, executive management includes it in the capital budget. Final approval came from the board of directors. Since that time, Ideal has systematized its procedures for deciding upon capital expenditures, making the process much more rigorous and placing greater controls on those expenditures.

This particular decision, however, circumvented the then loosely structured capital allocation decision process. Senior management took the lead, deciding on a massive coal utilization program to reduce fuel supply vulnerability. Because the Houston plant was part of this overall commitment, it did not have to meet the firm's required rate of return.

THE CHEMICAL INDUSTRY: E.I. Du Pont de Nemours & Company, Inc.

I. CHARACTERIZATION OF THE INDUSTRY

A. Function

The chemical industry consists of three major industrial subgroups--industrial organics, industrial inorganics, and alkalies and chlorine--which produce a wide range of products for a multiplicity of markets. Sales to final users, of which exports make up the largest segment, account for only 18.1 percent of total industry output. Since most sales are to firms for further processing and distributing, firms tend to compete on the basis of costs rather than on product identification. (The firm used for the case study, Du Pont, is an exception. It has developed a number of innovative, high profit products.) Approximately one-fifth of the industry's inputs are internally derived. Since most firms are not vertically integrated, they tend to purchase primary materials such as petroleum feedstocks and energy resources from other industries.

B. Energy Use

Chemical companies use energy for process heat, boiler fuel, and feedstock. Fuel shifts have run towards oil and electricity and away from natural gas in response to curtailments in supply. Because of their large steam and electricity needs, chemical firms may choose cogeneration projects when site requirements, contractual arrangements, and capital availability make it worthwhile.

Since 1977, the industry has reduced its dependence on natural gas from 47 to 37 percent by increasing its use of purchased electricity and petroleum based fuels. Coal use stagnated at around 9 percent, although coal use for boilers is relatively large. The 6.74 quads of energy materials purchased in 1978 make this the most energy-intensive of the major manufacturing industries.

The chemical industry has substantially reduced energy demand. Net consumption per unit of output was reduced by 20.9 percent from 1972 to 1979. Housekeeping measures and technological innovations in manufacturing processes have been significant factors in meeting these goals.

C. Economics

Capital intensity in the chemical industry is quite high and profitability depends on a high capacity utilization rate. Unlike some other commodity producers, such as the steel industry, the chemical industry is more protected from cyclical troughs by product diversity, and corporate plans tend to focus on shifting resources to account for the changes in various markets.

II. FUEL-USE PARAMETERS

A. Technical Barriers to Coal

Technical incompatibility between chemical processes and fuel burning processes poses problems for coal use only in selected instances. Coal provides fuel for process steam production and, to a lesser extent, for self-generated electricity. Product contamination through direct heating precludes a wider use for coal, but the potential for increased reliance on coal is great. The industry as a whole is in the process of reconverting coal-capable boilers now using oil or gas back to coal.

B. Fuel Costs and Supply Experience

A primary concern of chemical manufacturers is the reliability and security of the supply of energy resources. In response to fuel supply concerns, many firms retain oil storage equipment, even at plants that have been reconverted to coal. Capacity utilization necessitates a continuous supply of available energy sources for the ongoing manufacturing processes.

C. Environmental Regulatory Experience

Because of safety issues within plants, dramatically publicized episodes such as Love Canal, and government requirements, the industry has considerable experience and familiarity with environmental legislation, especially in the areas of air and water pollution. Pollution control expenditures accounted for 6 percent of capital expenditures in 1981. One firm estimates that total industry operation and maintenance costs for pollution control programs will reach $1 billion by 1985 (in 1985 dollars). So far spending has gone mostly towards compliance with air and water pollution control legislation, but the upcoming decade will see efforts to account for hazardous waste disposal under the Resource Conservation and Recovery Act. Most firms have standing committees on environmental policy, engage in intensive public relations efforts to improve their public image, and mount well-funded lobbying efforts to affect legislation.

III. ILLUSTRATIVE FUEL-USE DECISION

A. Background

The case study focused on a conversion proposal for a Du Pont neoprene plant, located in the gas-rich Texas-Louisiana area. That plant is currently powered by two oil/gas-fired 220 mmBtu/hr feed rate boilers. The boilers, which use gas, were not operated at anywhere near full capacity. The new facility would include one 240 mmBtu/hr pulverized coal-fired boiler operated at capacity with a baghouse unit, coal handling system, and auxiliary facilities. The plant is located in an attainment area for major pollutants (SO_2, NO_x, and TSP). Natural gas is the primary fuel source for the region.

B. Options Considered

Du Pont pursued the conversion project in response to an expiring natural gas contract and a company policy to explore coal potential whenever possible. Natural gas was being phased out because new contracts were considered too costly. Oil was rejected after a summary review because prices were projected to rise faster than those for coal. Cogeneration was not deemed economic for this site due to reasonable electric rates from the utility.

C. Major Decision Factors

The driving force in favor of conversion was Du Pont's assessment of the economics of coal use, which was based on a projected widening of the gap between coal and oil prices. Two incentives helped drive Du Pont's initial decision: (1) the presumed imminent promulgation of New Source Performance Standards for industrial boilers in the 100-250 mmBtu/hr range, and (2) scarcity of available Prevention of Significant

Deterioration increments in the region. If Du Pont could obtain its permits before the new standards were published and before competitive industrial development obtained the remainder of the clean air increments for the region, the conversion project would compare favorably with alternative fuel-use decisions.

Du Pont planned to use low sulfur eastern coal (1.0 average) to meet SO_2 control requirements. The baseline pollutant level for SO_2, by which PSD increments are measured, is low for this area since natural gas has been the primary fuel. A major oil company has recently obtained most of the allowable increments in the region and the chemical firm's application effectively secured the remainder.

The permitting procedure with the regional EPA office caused delays in the project. Du Pont was not able to work with the state environmental agency, which has good relations with industry, and instead had to deal with the regional EPA office. Du Pont was forced to do modeling and monitoring to determine the base from which the PSD increment would be calculated. An outside contractor to EPA, however, was dissatisfied with the firm's data and had requested more information. Once the permit was formally submitted to EPA, approval occurred within twelve months. Du Pont is unlikely, however, to move ahead with the project in today's economic climate.

D. Decisionmaking Process

Du Pont is a large multinational corporation with a complex organizational structure. Even though capital budgeting is constrained, just as it is for any other firm, the company enjoyed a low debt to equity ratio at the time of the study, and could, if it needed to, increase capital expenditures considerably. During the 1960s and 1970s coal use as a percentage of fuels for steam production dropped sharply, but the company now has a stated policy of reconversion to coal. It has exceeded other chemical manufacturers in reducing energy demand.

With a large, well-trained staff for energy and environmental analysis, Du Pont rarely uses consultants. Several groups within the corporation may be called in on questions of fuel use--a power consulting group, a materials and energy group, and engineering, finance, and environmental affairs departments. Usually, these groups are consulted on decisions involving large expenditures. Smaller energy projects may be approved by a departmental vice-president without entering the elaborate capital budgeting scheme.

Du Pont has a hierarchical decisionmaking structure, with capital proposals generally originating from plant managers and then wending their way through department and staff groups to the board of directors. The corporate plans department coordinates and ranks, according to long range strategic goals, all capital allocation requests, including large expenditures, for boiler conversions. All investments are ranked according to necessary, offensive, and defensive objectives. Executives develop plans for capital based on market conditions for various products. For example, a coal reconversion project at a plant involved in the manufacture of a product with a weak competitive stance or in a low growth market might be foregone even if cost savings are large. Nevertheless, Du Pont does have a corporate objective of reconverting 95 percent of its formerly coal-capable plants back to coal, and a policy to review coal use and cogeneration potential at all new facilities.

THE FOOD PROCESSING INDUSTRY: General Foods Corporation

I. INDUSTRY PROFILE

A. Function

The food industry (Standard Industrial Classification 20) consumed 7 percent of all purchased fuels and electricity used by manufacturers to process raw farm output into hundreds of food products, some for direct consumption and others requiring further processing. The diversity of output (hundreds of products in nine major subgroups) defies generalization. Although some of the ensuing discussion refers to the entire food industry, most of it focuses on cereal manufacturing, the subject of the case study.

B. Energy Use

Concentrated in the Midwest, most food processors rely heavily on natural gas (50 percent of all energy). Grocery manufacturers--a subset to which General Foods Corporation belongs--derive slightly more energy (60 percent) from natural gas.

Energy use in the industry has risen dramatically in recent years because of consumer preference for highly processed foods, transferring energy requirements from the home to the manufacturer. Yet the industry has an excellent conservation record, especially in the meat and canning subgroups. Efficiency improved by 17.4 percent in 1978 over the base year of 1972.

C. Economics

The ten industry subgroups may be classed under broader headings to describe basic economic influences. Commodity producers, such as sugar refining and wet corn milling, compete on the basis of price. Another class of producers is made up of small regional food packer/processors, which produce on contract for retail private labels or market their own regional brands.

The third class, the subject of the case study, is the packaged or branded food group. This class competes on the basis of consumer demand for a specific product, using large advertising expenditures to create a perceived difference and brand name identification in the consumer's mind. The industry depends on a high volume of product development, a constant turnover and change in the number of items available to the consumer, and a timely delivery of shipments to retailers, who try to minimize inventories. In cereal manufacture, four or five major firms dominate shelf space in retail grocery outlets. They are competitive in advertising but less so on price. Price increases for factors of production (e.g., fuel) represent a small percentage of total costs. Although competition centers on brand identification, some competitive pressures result from generic (no-name) products.

II. FUEL-USE PARAMETERS

A. Technical Barriers to Coal

Industry-wide, boilers for steam and electricity consume 51 percent of fuels purchased, while direct heat uses only 10 percent. But for the cereal foods manufacturer of the case study, 30 percent of energy is used for direct heat, an application that prohibits coal use because of product contamination. Even for boilers, coal use in the industry is restricted because most plants have boilers under 75 mmBtu/hr.

B. Fuel Costs and Supply Experience

With price controlled natural gas the basic industry fuel, food manufacturers have been largely shielded from the effect of rapid price increases. For cereal manufacturers, energy represents a low percentage of total costs, in the range of 1 to 2 percent. Natural gas is the preferred fuel for two reasons. First, it has been a cheaper and cleaner fuel to burn. Second, the Federal Energy Regulatory Commission's (FERC) natural gas priority classifications include the food industry under agricultural uses, which are exempt from curtailments. Thus, the industry is shielded from concerns about fuel availability.

C. Environmental Regulatory Experience

Natural gas dependence has kept most food manufacturers free from extensive dealings with pollution control agencies. Because the industry is also sensitive to consumer opinion, firms strive to maintain a clean and spotless image at the plant, which might be tarnished if "dirty" fuels were used.

III. ILLUSTRATIVE FUEL-USE DECISION

A. Background

General Foods is a large domestic food processor that markets grocery coffee products, packaged convenience foods, and food-away-from-home franchises. The plant, located in Battle Creek, Michigan, that formed the core of the case study used coal for steam production in manufacturing a variety of cereals. The plant has had multiple fuel capacity (coal, oil, gas) from three boilers, two of which burn oil, natural gas, or pulverized coal; the third burns oil or gas.

Plant operations are the most efficient in the firm because of sophisticated instrumentalization. Coal has been the dominant fuel source despite a sulfur emissions problem in the early 1970s, which was solved by switching to low sulfur coal.

B. Options Considered

General Foods was prompted into action in 1976 when the state pollution control agency notified the firm that particulate emissions from coal combustion exceeded the standard. The firm reviewed four different options generated by a consulting engineer to meet the compliance order. General Foods could: (1) improve existing mechanical collectors and continue to burn coal; (2) install baghouses at a cost of $2.5 million, 10 percent of total capital invested in the plant; (3) burn a 60:40 mixture of oil/coal; or (4) switch from coal to oil, gas, or a mixture. The last two options involved no capital investment.

C. Major Decision Factors

The first two options were eliminated quickly. The first was rejected because the improved collectors could not demonstrate sufficient control of particulates. The installation of the baghouse was considered too large an investment because of the age of the plant and remaining useful life.

As the decision was being evaluated, the state pollution control agency informed the company that the oil/coal option would not be acceptable, since particulate emissions must be measured separately for each fuel. The agency claimed that air quality could not be assured with the coal/oil mix because of the difficulty in detecting actual emissions and because of a lack of control over the fuel mix. State officials were also concerned about allowing an "exception" to its rules. The company was thus left with only one option to pursue, the oil/gas option. Because General Foods was able to negotiate a favorable contract for natural gas, it began to use gas exclusively.

D. Decisionmaking Process

General Food's organizational structure reflects its emphasis on product identification and development with some thirty vice presidents handling product functions. In addition, General Foods has one vice-president for purchasing materials and one for operations.

The energy strategy is planned quarterly by an energy group, which submits its plans to the vice-president of operations. Company-wide, this plan has resulted in the installation of dual oil and gas capabilities in all plants. The corporate strategy for energy planning takes into account relevant areas of information--fuel supplies and prices, regulatory or technological changes, and assumptions about the political and economic situation in the world. Because energy costs are relatively small (1.5 percent of product sales price), General Foods is more concerned about reliability and flexibility of supply than about costs.

THE PULP AND PAPER INDUSTRY: St. Regis Paper Company

I. INDUSTRY PROFILE

A. Function

Manufacturers of pulp and paper convert wood into a variety of intermediate products--graded pulps and wholesale paper stocks. For the most part, output of final retail products is left to converters and printers. Papermaking involves pulpwood acquisition, debarking, chipping, pulping by either chemical or mechanical means, bleaching, paper production, and converting. Typical mills are integrated facilities, but some firms purchase pulp and other materials.

B. Energy Use

The pulping and paper production stages together account for 89 percent of the energy requirements for the process. Mechanical pulping requires electric or mechanical energy to drive grinders; chemical pulping uses low temperature steam in a cooking process. The fabrication of paper and paperboard from pulp involves the formation of a sheet of fibers and the removal of excess water with steam or radiant heat.

Manufacturers use a variety of fuels. Bark wood wastes and spent pulping liquors comprise 50.8 percent of the industry's fuel use. Although oil and gas usage has decreased in recent years, those fuels still comprise the bulk of purchased energy resources. Over 51 percent of the industry's energy requirements are generated internally by hydroelectric generators and by cogeneration.

Because of its reliance on self-generated fuels, the pulp and paper industry reports its conservation efforts differently from other manufacturers in terms of purchased energy rather than total energy per ton (.907 metric tons) of product. On this basis, the average use of energy per ton (.907 metric tons) declined by 17.8 percent during the 1972-79 period. Overall efficiency proceeded more slowly, since waste fuels that the industry is using to replace fossil fuels are less thermally efficient; total energy per ton (.907 metric tons) declined by 7.2 percent. The major economies were achieved through the replacement of small plants with integrated mills and the increased use of by-products as fuel.

C. Economics

Demand for the industry's end products is cyclical and tends to track changes in business conditions. Profitability depends on a high capacity utilization rate. Paper firms compete on the basis of price by specializing in one of the many submarkets such as newsprint, printing papers, or liner board. In order to capture new market shares within subgroups, and because of general scale economies, paper firms rely on substantial capacity additions in anticipation of demand growth. For these reasons, paper companies have traditionally had to commit large sums of capital during troughs in the industrial cycle.

Over the years, paper manufacturers have experimented with a number of structural strategies--

unrelated diversification, backwards integration to assure access to raw materials, and lateral moves toward other forest products. The industry's low profit margin firms have increased their long-term debt as a percentage of total capitalization. These problems with capital availability impinge on the prospects for the increased use of coal.

II. FUEL-USE PARAMETERS

A. Technical Barriers to Coal

Many of the newer boilers for the industry were designed especially for oil or natural gas and these still have many years of useful life. Industry representatives also are concerned about their ability to obtain state land use permits for greenfield plants. Development of existing sites through modernization and expansions means considerable spatial constraints. Fugitive dust from coal storage might pose problems in the contamination of products awaiting shipment. Large steam requirements, however, suggest that, from a technical point of view, coal has substantial potential.

B. Fuel Costs and Supply Experience

The pulp and paper industry is the largest industrial user of fuel oil and has seen enormous increases in energy costs since 1979. Despite this experience, most firms have been reluctant to forecast fuel prices and have generally made conservative assumptions about price differentials between oil and coal. Some see coal prices catching up to oil prices and express concern over bottlenecks in coal supply.

C. Environmental Regulatory Experience

The industry is familiar with the provisions of environmental regulations and has already made large investments to comply with them. Nearly two-thirds of pollution control expenditures to date have gone toward control of water pollution. Since many firms are located near Class I regions, air pollution concerns center around PSD modeling of air quality and regulations regarding visibility.

Energy costs are large and at least some factor in competition. The high percentage of energy costs per unit of output has increased the attractiveness of coal or wood options for many firms. The key limiting factors are site constraints, pressures to make productive investments, transportation costs, and fear of product contamination.

III. ILLUSTRATIVE FUEL-USE DECISION

A. Background

St. Regis Paper Company is a large manufacturer engaged in various segments of the industry including pulp, newsprint, printing papers, dielectric papers, kraft paper and board, and construction and industrial packaging papers. It also manufactures packaging machinery and systems, lumber, plywood, and other construction materials.

The mill chosen for study is a semi-integrated pulp and paper mill in the Northeast that produces coated paper for magazines. Its output reflects a major share of this industry submarket. Demand for the product is strong and has allowed the plant, which operates at capacity, to expand during the 1970s. The plant, located in a Class I region for the attainment of NAAQS, has thirty acres of usable space providing for the plant buildings, a wood yard, a disposal site, a water treatment facility, and a rail spur.

The power plant for the mill consists of one large and two small oil-fired boilers, which replaced four coal burning boilers. The coal boilers, still in place, cannot be used because their age renders them uninsurable. The current boilers burn 2.5 percent sulfur residual oil. Between 1979 and 1980 fuel costs rose substantially for the plant.

B. Options Considered

In order to eliminiate its oil dependency, St. Regis considered a number of options suggested by its technical consultants to meet its energy requirements. Three conversion possibilities were seriously considered--coal, wood, and a coal/wood mixture. Other alternatives, which were dismissed after initial review, included additional cogeneration capacity, coal gasification, burning of town refuse, and the use of peat.

C. Major Decision Factors

The coal option initially appeared less promising because it entailed higher fuel costs and a higher capital investment to meet pollution control requirements. The firm initially favored the all wood option because it eliminated the SO_2 control requirement and could contribute, through the use of debarking wastes, to a reduction in the plant's solid waste disposal problem. In addition, the company would have more control over fuel supplies since it owns considerable forests, including hardwood not appropriate for pulping.

Two factors seriously constrained the decision on any of the options. The site was not easily expandable and investments initially fell short of the corporation's return on investment (ROI) goal. The company had decided to forego conversion plans but subsequently changed its policy in response to rapidly escalating oil prices. In 1981, the firm decided to pursue the coal/wood option because of the riskiness of oil supply and an improved projected return on investment.

The coal/wood option was substituted for the initial plan for several reasons. First, technical personnel discovered that a completely wood-fired system would be unable to produce the surge power needed intermittently and that some oil use would still be necessary. Second, managers were concerned that local supplies of hardwood might not be available without serious price impacts. The coal/wood mixture also allows for some flexibility in fuel use, which is not generated under an all wood or all coal option.

St. Regis feels it is able to comply with the strict standards of the state environmental agency. PSD increments are scarce in the region and there is competition from a local utility for their use. The state must also grant approval for site development based on neighboring land use. Solid waste disposal is the most difficult and expensive permit to obtain because of the land requirements. Nevertheless, a senior management policy to remove the firm's oil dependency is forceful enough to drive the fuel-use decision towards implementation. The company expected the permitting process to take one year.

D. Decisionmaking Process

Managers at all levels of the corporation participate in energy decisions. In the last several years, the firm has moved towards the centralization of energy decisions by establishing a corporate energy committee, which assesses overall performance, and a subcommittee, which investigates site specific opportunities. Before 1979, fuel purchasing was assigned to various individuals by type of fuel. This authority is now vested in a manager of corporate energy purchases.

The corporate planning department together with the energy committee has evolved a policy for switching away from the use of oil. The firm assumes that current fuel price differentials will hold in the next few years and sees problems with the security of oil supply.

THE PETROLEUM REFINERY INDUSTRY: Exxon Corporation

I. INDUSTRY PROFILE

A. Function

The refinery industry is comprised of a range of producers who make several hundred grades of products in more than a dozen product categories. More than 80 percent of refinery outputs are final energy products such as gasoline, jet fuel, distillate, and residual fuel oils sold to end-users. The remaining outputs are intermediate products like petrochemical feedstocks used as inputs to other industrial processes or non-energy products such as asphalt or lubricating oils.

B. Energy Use

The industry uses 70 percent of its process energy in direct heat processing; steam and electricity requirements account for the remainder. To fuel these operations, a number of energy sources, most of them internally generated, are used, but the major sources are refinery gas, natural gas, petroleum coke, and electricity. Coal's share of the fuel mix has been very low. Purchased steam makes up just over 1 percent of the fuel mix.

Conservation measures between 1972 and 1979 allowed total energy consumption to rise 3 percent while refinery throughput climbed 15 percent. However, the American Petroleum Institute (API) claims that these figures undervalue efficiencies because more energy is used in the production of unleaded gasoline, desulfurization, and pollution control.

C. Economics

Demand for petroleum products is rapidly shifting. Since late 1980, total petroleum demand has been falling. Demand for lighter products, particularly distillate fuel oil and diesel fuel, are rising, while demand for residual fuel oil has declined. In addition, refiners will be forced to use heavier crude oil supplies. These shifts have required refiners to make substantial investments to upgrade refineries to produce more light products from heavier crude oil supplies. With the elimination of crude oil price controls and the small refiner bias, a number of small, inefficient refiners--producing predominantly heavy fuel oil--have gone out of business. There are reasons to believe that discretionary investments in coal conversion are not likely--at least until economic conditions improve. Oil company profits have been in a slump since real crude oil prices began declining in 1981. The refinery segment of the industry, not historically the most profitable segment, has been allocated large amounts of capital for upgrading plants; it is hard to believe that a large new infusion of capital would be available for coal conversions. With the cash flow available, most oil companies will place the highest priority on finding and developing new supplies of oil.

II. FUEL-USE PARAMETERS--COAL-USE POTENTIAL

A. Technical Barriers

Contamination is not a concern of refineries since combustion does not occur in direct contact with products. The large amount of energy inputs in the form of steam and electricity for heat treating and machinery are sufficient to justify coal cogeneration investments. Cogeneration is feasible if the system balances steam pressure and volume with electrical requirements.

B. Fuel Costs and Supply Experience

Fuel costs represented approximately 7 percent of the value of shipments in 1979. Although the refinery industry predominately relies on by-product fuels, it must purchase some additional energy. If coal could replace natural gas or oil for generating steam, the oil or gas could be sold.

C. Environmental Regulatory Experience

From 1969 to 1978, refinery capital expenditures for air pollution totaled more than $1.5 billion, approximately 75 percent of total expenditures for the entire petroleum industry. The American Petroleum Institute has detailed several project delays and cancellations because of difficulties implementing PSD and offset regulations. Refiners complain about the lack of adequate agency manpower to process permit applications and about policy conflicts between state and federal regulators.

III. ILLUSTRATIVE FUEL-USE DECISION

A. Plant Background

The case study focused on Exxon Corporation's Baton Rouge refinery. The plant is in an attainment area for particulates and sulfur dioxide and in a nonattainment area for ozone. The refinery receives its steam and electricity supply from a utility. Concerns about reliability because of age and capacity factors at the utility led the refinery to embark on a bold project to consider building four mammoth coal- fired boilers (500-800 mmBtu/hr) to supply all steam and electrical requirements.

B. Options Considered

Exxon considered a wide range of options before tentatively deciding on a coal-fired cogeneration system to produce steam and electricity. At the time of the case study, the decision was pending final design specifications. This assessment alone will cost $200-300 million and the total project cost is now estimated at $2 billion. Because of current market conditions in the petroleum industry, the Baton Rouge coal cogeneration project has been deferred.

C. Major Decision Factors

At the time of the case study, Exxon was projecting increases in world oil prices, ultimately reaching the level of synthetic fuels costs. Since it was projecting only modest increases in coal prices, the advantages of using coal would increase substantially over time. Moreover, the substantial economies of scale for the project could lead to favorable economics. The project was evaluated as part of the firm's formal planning process, with both headquarters and field staff participating in evaluation and recommendation to the Board.

THE STEEL INDUSTRY: United States Steel Corporation

I. INDUSTRY PROFILE*

A. Function

Steel production--which involves coke production, iron production, refining, and product formation--frequently occurs at one location in an integrated mill. Each stage is fully related to other steps in the process, since initial by-product fuels are used to supply energy requirements for later steps.

Coke production--anaerobic heating of coal in batteries of coke ovens--uses 11.3 percent of all

* The steel industry is currently undergoing major structural changes. The information presented here is based on industry data for 1978.

energy consumed in steel production. This stage also provides energy by the release of by-product gases and liquids. Approximately 40 percent of the coke oven gas fires the coke ovens, with the remainder used in steelmaking operations. More than 95 percent of coal consumption in steelmaking goes into the formation of coke.

The second step, iron production, entails the smelting of coke and iron in a blast furnace. This process is fueled by coke supplemented by natural gas, fuel oil, or coal tar. Blast furnace gas is a by-product used to preheat air for the furnaces. Iron production consumes 60.5 percent of all energy in steelmaking.

Actual production of molten steel uses only 6.7 percent of total energy to blend iron and scrap steel and to refine them in one of three types of furnace: basic oxygen, open hearth, or electric arc. The basic oxygen furnace requires no direct process fuel; the open hearth furnace may be fired by a range of fuels including by-product gases, oil, or natural gas; electric arc furnaces run entirely on electricity either purchased or internally generated.

Finally, steel product formation again requires energy to heat raw steel for shaping. These processes use 21.5 percent of the total energy consumed in steel production. Newer mills employ continuous casting techniques that shape end products directly from molten steel and avoid the reheating of primary ingots. Concerns about product contamination make coal use unfeasible at this stage of the process. Energy is derived from remaining coke oven and blast furnace gases supplemented by natural gas or fuel oil.

B. Energy Use

Coal's contribution to steel production has declined slightly since 1972 (from 69 to 65 percent of total energy use) due primarily to a drop in coke production. Some coke production has been foregone because of efficiency improvements in blast furnaces, but much of the coke production has been replaced by imports. The decline has also resulted in an increase in oil consumption in steelmaking operations to replace lost production of by-product fuels. Electricity use has increased slightly and is expected to grow as modernizations favor electric arc furnaces.

Since energy costs have always been a significant factor in production costs, the industry's conservation efforts antedate those of most other manufacturers. Between 1950 and 1978, average consumption of delivered Btu per ton (.907 metric tons) of raw steel produced declined 28 percent; only 9 percent of this reduction, however, has occurred since the usual benchmark year of

1972. Reduced coke consumption in blast furnaces, resulting in improved efficiency, has influenced this decline; another factor is the reliance on the basic oxygen method to produce over 56 percent of total production.

C. Economics

Steelmaking is a basic product industry subject to cyclical demand. Over ninety domestic companies produce steel but the industry has a concentrated ownership structure. In 1978, the top six firms were responsible for 64 percent of raw steel production. Slow GNP growth over the last decade, declining markets, especially in the automobile and construction industries, and tough foreign competition have imposed severe restraints on capital availability. Substantial capital investment will be required if U.S. firms are to keep pace with the high productivity rates and low prices of their foreign counterparts. But low profits reduce the cash flow available to make such investments. Imports of steel accounted for over 18 percent of U.S. consumption in 1978. The estimated useful life of equipment for U.S. mills is twenty-five to thirty years, with current stock having an average age of seventeen and one-half years.

External debt as a percentage of capital investment has risen from 28 percent in 1974 to 36 percent in 1978. Substantial inflation in the costs of new equipment puts added strain on capital budgets. Internal financing from retained earnings has suffered from a decline in after-tax profits as a percentage of stockholders equity. In the past, steel revenues have been held down by the efforts of several federal administrations to keep steel prices low in order to reduce inflationary pressures in the economy.

Replacement of existing plant and equipment has been slow because of these difficulties. Much of the steel industry's investment, however, has the character of necessary expenditures. Mandated pollution control investments make up 15 to 20 percent of capital spending, but they are no less binding on firms than replacement of old equipment.

II. FUEL-USE PARAMETERS

A. Technical Barriers to Coal

Since coke production/consumption represents the basic avenue for coal consumption in steel production, much will depend on the replacement of aging coke batteries. The greatest barrier to increased coke production would appear to be the high capital requirements [$200 million for 1 million tons (907,000 metric tons) of annual capacity]. Pollution control costs represent about 20 percent of the total capital costs. A new technology to enable the direct use of coal in blast furnaces has been developed but is still largely uneconomic. The industry also uses numerous boilers; however, most of this capacity already uses coal or has been ruled out for conversion due to boiler size. The downstream processes (steelmaking and forming furnaces) are restricted to clean fuels because of product contamination concerns.

B. Fuel Costs and Supply Experience

Fuel costs as a percentage of total costs for the steel industry have risen steeply in recent years, from 12 percent in 1972 to 22 percent in 1978. This reflects an average annual rate of increase of 20 percent. Although natural gas interruptions have not been a widespread occurrence, diminishing natural gas supplies during the early 1970s encouraged most steel producers to maintain a fuel oil stockpile. Many industry members emphasize fuel-use flexibility as the best protection against potential problems.

C. Environmental Regulatory Experience

Environmental regulations have had a bearing on energy efficiency. The declining quality of the coal available for metallurgical purposes requires more energy to remove sulfur and ash. Environmental controls account for 2 percent of the energy required to produce steel and will consume 7 percent by the mid-1980s, according to the American Iron and Steel Institute. Although steel manufacturers have already spent $4 billion since the 1950s on pollution control hardware and will spend that much again within the next five years, the industry has faced compliance problems. An adversarial relationship has developed between many firms and federal regulators over issues concerning the uncertainty of requirements and conflicting decisions by overlapping environmental agencies. These difficulties are exacerbated by the economic problems faced by steel manufacturers trying to control declining rates of profitability.

III. ILLUSTRATIVE FUEL-USE DECISION

A. Background

The United States Steel Corporation's plant, located in Pennsylvania, was forced into a fuel-use change by a state issued environmental order to bring its boilers into compliance to meet particulate emission requirements. Two of the boilers at the facility serve as the focus for this decision.

B. Options Considered

Two boilers, which were adapted to use a variety of fuels, provide steam to drive the blast furnace turbo blowers and to help generate 7 percent of the plant's annual electricity requirements. At the time of the state order, the boilers operated on the following fuel mix:

blast furnace gas	64%
coke oven gas	27%
coal	6%
natural gas	2%
fuel oil	1%

The firm's original decision was to divert some of the coke oven gas for other operations, move away from natural gas because of concerns over curtailments, and to use more coal or oil. Oil use would require capital outlays for storage and for steam for heating residual fuel oil. Increased coal combustion required capital expenditures for precipitators.

C. Major Decision Factors

Initial cost studies favored oil use but proved inaccurate because of an underestimate of capital costs. With a new estimate in hand, U.S. Steel decided on coal, due to a favorable low cost fuel mix of coal, coke oven gas, and blast furnace gas. The boilers now use 46 percent mixed gas (coke oven and natural gas) and 54 percent coal as the result of the shut down of blast furnace operations in 1982.

Regulatory problems did not prove to be an issue since the firm has a good working relationship with local officials and since sulfur emissions problems were not in question; extensive reliance on by-product fuels kept sulfur emissions low.

Since the project required capital expenditures of over $7 million, it was reviewed under corporate budgeting procedures. Some decision was necessary because of the compliance schedule dictated by the air pollution agency. But the allocation of funds for precipitators at this specific location reflects the firm's interest in the plant as an efficient and competitive facility. Several other U.S. Steel mills were closed in 1979 because capital investments were too costly for them.

An environmental requirement defined only one of the parameters for the decision in this case. The final solution was primarily a factor of the steel-making process itself and the company's need to optimize its fuel mix according to the many energy requirements at the plant. Since the industry by nature is a heavy user of coal, it finds multiple uses for an already available resource. The prospects for an increased reliance on coal will depend on the potential for added coke production and on the ability of the industry to overhaul its plants and regain its competitive position in world markets.

D. Decisionmaking Process

In 1979, U.S. Steel suffered severe financial losses, especially in its steel manufacturing division, and the consequent closing or reduction of a number of facilities. The company's other divisions, including chemicals, resource development, and domestic transportation and utility subsidiaries, did well and received increased attention and allocation of funds. The steelmaking division has to compete with these other segments for capital funding.

The steel division is divided regionally with separate functional departments managed by vice-presidents. Divisional managers focus directly on higher profits; individual plants operate entirely as cost centers. This separation is reflected in the levels of sign off authority that each is allowed. Plant managers, assisted by their own power and fuels group, decide on day-to-day fuel mixes and marginal shifts in supplies. They can redistribute fuels such as coke oven gas or oil from a blast furnace to a finishing operation. A limit of $100,000 to make energy efficiency improvements is allocated to them. Divisional managers allocate fuels across plants and develop new ideas about energy use and technology. The divisional vice-president has a sign off authority of $7 million and can accommodate a range of fuel-use changes.

An energy committee assists corporate-level planning and acts as the major conduit for response to the government's conversion program. A capital budgeting system, where major projects enter a pool of other competing investments, becomes part of a tactical plan for the firm's near term strategies. These are continually revised or altered according to circumstances. Energy decisions at this level do not get preferential treatment, but they do receive careful attention. The firm conservatively expected the world price of oil to rise at least with the rate of inflation.

A SMALL PAPER MANUFACTURER*: Johnson Paper Company†

I. CHARACTERIZATION OF THE FIRM AND ITS FUEL SWITCHING POTENTIAL

A. Function and Economics

The firm, located in the Northeast, is a small, family-owned, producer of a specialty paper product, carbonizing tissue. It employs 100 workers and has an annual gross sales level of $10 million. An anomaly in an industry characterized by large operations that take advantage of economies of scale, the firm must struggle to remain cost competitive in its single product niche.

B. Energy Use

As a paper manufacturer not engaged in pulp production, the firm uses a combination of electricity (2 megawatts, average demand) to drive the machinery and steam (15,000 to 23,000 kpph, depending on the season) to heat the paper driers. Prior to its fuel conversion, the company purchased all its electricity and fired its boilers with #6 fuel oil. The firm produces no pulping by-products that could be burned as fuel.

The principal motivating factor in the firm's decision to investigate alternative fuel possibilities was the steep increase in fuel oil prices experienced in 1979, particularly relative to its competitors. Required to use low sulfur fuel oil by the State Implementation Plan, the company bore a doubling in fuel oil costs in one year. Most of its competitors, due to their size or location, could use higher sulfur fuel oil. The premium, which the firm was forced to pay, rose from $3 to more than $5 per barrel. In a price competitive industry, the firm had to act to preserve its profitability.

C. Environmental Regulatory Experience

The firm has successfully handled its water quality requirements with the installation almost ten years ago of its own water treatment plant to process both incoming and effluent water.

---

* Because the industry setting for this case study is described in some depth in the pulp and paper case study summary, the introductory section below will be confined to an effort to place the small firm in the larger context.

† At the firm's request, Johnson Paper is a pseudonym.

II. ILLUSTRATIVE FUEL-USE DECISION

A. Options Considered

Faced with the compelling need to bring energy costs under control, the company examined fuel switching for its boiler and also cogeneration to reduce electricity costs. In the case of the former, the company reviewed two options, coal or wood, but initially abandoned the latter based on fuel availability and cost projections. Since all indicators pointed favorably to coal, the firms decided to go ahead with the conversion. A last minute breakdown in financing arrangements forced the firm to abandon coal and turn to wood instead. Had the financing come through, however, the firm would have converted to coal.

The coal project was to include the boiler system, pollution control systems, turbines for cogeneration, and a building enclosure to reduce fugitive dust and particulate emissions. To insure reliability, the company was to maintain a ten day supply of both coal and oil and use existing boilers as a backup system. The use of low sulfur coal, a baghouse, and a building enclosure would reduce concerns over permitting for the conversion project.

B. Major Decision Factors

The company based its decision largely on an evaluation of the projected fuel cost savings from conversion and cogeneration investments. An outside contractor performed the technical and financial studies needed to assess the worth of the project. His conclusions, based on an increased differential between the prices for oil and coal, pointed to substantial savings for the firm. The contractors estimated capital costs to run just over $2 million with operating savings due to coal conversion expected to be over $500,000. The capital expenditures for the conversion would increase the firm's assets by 70 percent. The savings attributed to cogeneration during 1982 were estimated at $159,000. This assumed electric power would cost 9¢/kwh. Cogeneration seemed especially attractive because the added capital investment for turbines was only $200,000, merely 9.4 percent of total expenditures.

The conversion project might have been unattractive if the company had not been able to factor in tax exempt financing for the portion of the investment directly related to pollution controls. The firm was able to secure permission from the state industrial finance authority to issue Industrial Revenue Bonds (IRBs), bonds whose interest is exempt from federal taxation. Half the value of the bonds was to be guaranteed by the Small Business Administration. Also crucial to the financial success of the company's plans was obtaining an environmental variance to burn high sulfur oil while the conversion proceeded. The use of cheaper, high sulfur fuel oil used during the capital investment period would generate cash flow that could be applied to the conversion. Although the permitting process ultimately ended favorably, a brief review of its duration provides a lesson in the difficulties smaller firms face in fuel switching.

The variance application required approval from a number of levels within the state environmental quality agency and from both EPA headquarters and the regional office. The director of the state air quality control region where the firm is located supported the application because it was well documented and showed good faith in working towards state and national energy objectives. Since the variance applied only during the time needed for conversion, air quality would not be seriously affected over the long-term. Approval would result, however, in the consumption of a large share of available PSD increments for the area. Quick and close relays of information between the state agency and the firm helped to expedite earlier reviews, but the final approval was delayed by EPA concerns about acid rain problems arising from decisions to permit higher sulfur emissions.

C. Decisionmaking Process

The corporate decisionmaking process involved only two individuals, the company's president and vice-president, members of the family who own the business. The crucial analyses of the project's economic benefits were performed by a consultant.

A SMALL ROSE GROWER: Montgomery Rose Company, Inc.

I. INDUSTRY PROFILE

A. Function and Economics

The domestic rose industry provides roses for sale as fresh cut flowers and for use in ceremonial and decorative displays (wreaths, bouquets, etc.). The extreme perishability of the product (one week maximum with the use of preservatives) eliminates the use of inventories to even out the supply/demand cycle.

The most difficult problem for the industry is matching peak production, which naturally occurs in the summer, to surges in demand, which arise around the winter and spring holidays. Successful breeding and cultivation techniques have improved the match somewhat. Rose growers must achieve high profits during the brief periods of high demand to cover high annual fixed costs. In 1978 imports targeted at peak demand periods (March, May, November, and December) contributed to a severe decline in profits.

B. Energy Use and Fuel Cost Experience

Because markets can be as important as climate in locating rose growing operations, many domestic producers are in cold-climate states. Most of the fuel consumption--75 percent of average annual use--is required to provide constant heat to greenhouses during winter months. The fragility of the product also requires occasional rapid boiler fire-up to protect a crop during sudden early or late frosts. The greatest technical difficulty greenhouses face with the use of solid fuels is the ability to fire boilers quickly to meet these small, temporary heating demands, which may occur in non-winter months.

Fuel costs are an important and growing component (up from 16.4 percent in 1976 to 19.1 percent in 1979) of total greenhouse expenses. Naturally, these increases have hit eastern U.S. growers the hardest, the same members of the industry most plagued by imports. Rose growers are also concerned with the security of fuel supplies, since even a short interruption in heating can result in a crop disaster.

II. ILLUSTRATIVE FUEL-USE DECISION

A. Background

Montgomery Rose Company, Inc. is a medium-sized rose grower (210,000 rose plants in nineteen greenhouses) with annual sales of $1.5 million. Located in Massachusetts, it has experienced pressures from both imports and rising fuel costs, squeezing its before-tax profitability from 12 percent in 1977 to 1 percent in 1979.

B. Options Considered

Montgomery Rose burned coal for fifty years but converted to oil in 1959. As early as 1974, the family owner-operators became concerned about the cost of oil and began to consider converting to wood. Average annual fuel use equaled 750,000 gallons (2,838,750 liters) of residual fuel oil, 75 percent of which was burned in the period mid-November through March. To protect against a supply cutoff, the firm maintained a 200,000 gallon (757,000 liters) reserve.

C. Major Decision Factors

The availability of scrapped boilers from a local college was a major factor in the decision to burn wood. The firm paid only for the cost of moving the boilers. If new boilers had been purchased, they would have cost $200,000 each, which would have been too expensive to permit a decent return on the investment. The firm financed the entire $750,000 of conversion costs internally, relying on greenhouse personnel for much of the labor. The conversion increased annual operating costs because of the need for labor to operate the equipment twenty-four hours per day. Although original fuel savings estimates of $108,000 per year offered an acceptable return on the investment, a rapid widening in the gap between oil and wood costs (from the 1977 ratio of 2.5:1 to the 1980 ratio of 5:1) has improved the project's economic return.

The biggest problem in making the conversion arose from the procedures required to gain the environmental permit. Indeed, Montgomery Rose might not have pursued conversion had it been aware of the regulatory problems. The firm failed to apply for a permit prior to initiating construction, resulting in several violation notices and subsequent months of downtime after the plant began wood combustion. Montgomery Rose had difficulties in filling out forms and objected to the required test burn as prohibitively expensive for small firms.

D. Decisionmaking Process

As with many small firms, the decisionmaking process was simple and brief. The owner-operators decided to reduce fuel costs by taking advantage of locally available, cheap boilers. The firm lacked the experience to understand the requirements necessary for obtaining an environmental permit, resulting in delays and frustration.

Index